SLEIGHTS OF MIND

SLEIGHTS OF MIND

WHAT THE NEUROSCIENCE *of* MAGIC REVEALS ABOUT OUR EVERYDAY DECEPTIONS

Stephen L. Macknik
and Susana Martinez-Conde
with Sandra Blakeslee

HENRY HOLT AND COMPANY
NEW YORK

Henry Holt and Company, LLC
Publishers since 1866
175 Fifth Avenue
New York, New York 10010
www.henryholt.com

Henry Holt® and ® are registered trademarks of
Henry Holt and Company, LLC.

Distributed in Canada by H. B. Fenn and Company Ltd.

Library of Congress Cataloging-in-Publication Data

Macknik, Stephen.
Sleights of mind : what the neuroscience of magic reveals about our everyday deceptions / Stephen Macknik and Susana Martinez-Conde ; with Sandra Blakeslee. — 1st ed.
p. cm.
Includes bibliographical references and index.
ISBN 978-0-8050-9281-3
1. Optical illusions. 2. Magic tricks. 3. Neurosciences. I. Martinez-Conde, S. (Susana) II. Blakeslee, Sandra. III. Title.
QP495.M33 2010
152.14'8—dc22 2010030116

Henry Holt books are available for special promotions and premiums.
For details contact: Director, Special Markets.

First Edition 2010

Designed by Meryl Sussman Levavi

Printed in the United States of America

1 3 5 7 9 10 8 6 4 2

To our wonderful children,
Iago and Brais.
Thank you for all the magic.

CONTENTS

SLEIGHTS OF MIND

INTRODUCTION

CLARKE'S THIRD LAW: *"Any sufficiently advanced technology is indistinguishable from magic."*

NIVEN'S LAW: *"Any sufficiently advanced magic is indistinguishable from technology."*

AGATHA HETERODYNE ("GIRL GENIUS") PARAPHRASE OF NIVEN'S LAW: *"Any sufficiently analyzed magic is indistinguishable from science!"*

Have you ever wondered how magic effects work? Coins materialize out of thin air. Cards move through a deck as if pulled by an invisible force. Beautiful women are cut in half. Spoons bend. Fish, elephants, even the Statue of Liberty disappear before your eyes. How does a mentalist actually read your mind? How can you not see the gorilla in the room? Really, how can someone catch a bullet in his teeth? *How do they do it?*

Don't bother to ask a conjurer. When joining an organization of professional magicians, the initiate may be asked to take an oath: "As a magician I promise never to reveal the secret of any illusion to a nonmagician, unless that person also swears to uphold the magicians' oath. I promise never to perform any illusion for any nonmagician without first practicing the effect until I can do it well enough to maintain the illusion of magic." It is a code. A brotherhood. The magician who breaks this code risks being blackballed by his fellow magicians.

So what are we, a couple of muggles, doing writing a book on magic? Zipped lips aside, hasn't most everything about magic been revealed? Enter "magic" in the Amazon Books search box and 75,000 results pop up. Log in to YouTube and you can see just about every magic trick ever devised—often demonstrated by darling seven-year-olds in their bedrooms with Mom or Dad wielding the videocam. Visit Craigslist and choose from myriad charming descriptions of local amateur magicians. What's left to say?

Actually, plenty. This is the first book ever written on the neuroscience of magic, or, if you will, *neuromagic*, a term we coined as we began our travels in the world of magic.* Much has been said about the history of magic, tricks of the trade, the latest props, and psychological responses to magical effects. But neuroscience probes more deeply. We want to pop the hood on your brain as you are suckered in by sleights of hand. We want to explain at a fundamental level why you are so thoroughly vulnerable to sleights of mind. We want you to see how deception is part and parcel of being human. That we deceive each other all the time. And that we survive better and use fewer brain resources while doing so because of the way our brains produce attention.

Like so much that happens in science, we fell into magic by accident. We are neuroscientists at the Barrow Neurological Institute in Phoenix, Arizona. The BNI is the oldest stand-alone neurological institute in the United States and currently the largest neurosurgical service in North America, performing more than six thousand crani-

*Devin Powell, a writer for the popular science magazine *New Scientist*, described our early studies in a 2008 article that introduced the term "magicology" (the scientific study of magic) as an alternative to "neuromagic" (the neuroscientific study of magic). Although "neuromagic" is somewhat narrower than "magicology," both terms are roughly equivalent and usually interchangeable.

otomies per year. Each of us runs a research laboratory in the institute. Stephen is director of the laboratory of behavioral neurophysiology. Susana is director of the laboratory of visual neuroscience. Incidentally, we are married. Both of us are primarily interested in how the brain, as a device that is made up of individual cells called neurons, can produce awareness, the feeling of our first-person experience.* Somehow, when neurons are hooked up to each other in specific circuits, awareness is achieved. It's the ultimate scientific question, and neuroscience is on the verge of answering it.

Our foray into illusions began a decade ago when, as young scientists seeking to make a name for ourselves, we tried to rustle up some public enthusiasm for our specialty of visual neuroscience. In 2005, after accepting faculty appointments at BNI, we organized the annual meeting of the European Conference on Visual Perception, which was held in Susana's hometown of A Coruña, Spain. We wanted to showcase visual science in a new way that would intrigue the public and the media. We were fascinated with how science can explain something about the visual arts—for example, Margaret Livingstone's work on why the Mona Lisa's smile is so ineffably enigmatic. We also knew that visual illusions are fundamentally important to understanding how the brain turns raw visual information into perception.

The idea we came up with was simple: we would create the Best Illusion of the Year contest. We asked the scientific and artistic communities to contribute new visual illusions and received more than seventy entries. The audience (a mixture of scientists, artists, and the public) viewed the ten best illusions and then chose the top three. The contest, now in its seventh year, has been a huge success. Our Internet audience doubles every year, and our Web site (http://illusionoftheyear.com) currently has about 5 million page views each year.

Because of our success with the illusion contest, the Association for the Scientific Study of Consciousness asked us to chair its 2007 annual conference. The ASSC is a society of neuroscientists,

*Throughout this book we use the terms "awareness" and "consciousness" as synonyms.

psychologists, and philosophers united in the aim to understand how conscious experience emerges from the interactions of mindless, individually nonconscious brain cells.

As our opening move, we proposed holding the conference in our hometown of Phoenix, but the association's board nixed that right away because the city is an inferno midyear. Instead, they suggested . . . Las Vegas. Hmmm. Las Vegas is every bit as blisteringly hot in June as Phoenix, and if you take the lap dancing, gambling, and showgirls into account it is probably several degrees hotter due to friction. So apparently our colleagues in consciousness studies were looking for a bit of real excitement to spice up their thought experiments.

So Vegas it was. We flew there in October 2005 to do some scouting. On the flight over we asked ourselves: How could we raise the visibility of consciousness research to the public? We didn't want to do another contest. The answer began to germinate the moment our plane dipped its wings on approach to the Las Vegas airport. Out the window we could see, all at once, the Statue of Liberty, the Eiffel Tower, an erupting volcano, the Space Needle, the Sphinx, Camelot, and the Great Pyramid. Soon we were driving up and down the Strip, checking out hotels for our meeting space. We passed Aladdin's castle, the Grand Canal of Venice, and Treasure Island. It seemed too strange to be real. Then, bingo: the theme for our conference appeared. Festooned on billboards, taxicabs, and buses were huge images of magicians: Penn & Teller, Criss Angel, Mac King, Lance Burton, David Copperfield. They stared out at us with mischievous eyes and beguiling smiles. And then it hit us that these tricksters were like scientists from Bizarro World—doppelgängers who had outpaced us real scientists in their understanding of attention and awareness and had flippantly applied it to the arts of entertainment, pickpocketing, mentalism, and bamboozlement (as well as to unique and unsettling patterns of facial hair).

We knew as vision scientists that artists have made important discoveries about the visual system for hundreds of years, and visual neuroscience has gained a great deal of knowledge about the brain by studying their techniques and ideas about perception. It was painters rather than scientists who first worked out the rules of visual perspective and occlusion, in order to make pigments on a flat canvas seem like a beautiful landscape rich in depth. We realized now that magicians were just a different kind of artist: instead of form and color, they manipulated attention and cognition.

Magicians basically do cognitive science experiments for audiences all night long, and they may be even more effective than we scientists are in the lab. Now, before our in-boxes fill up with flames from angry colleagues, let us explain. Cognitive neuroscience experiments are strongly susceptible to the state of the observer. If the experimental subject knows what the experiment is about, or is able to guess it, or sometimes even if she incorrectly thinks she has figured it out, the data are often corrupted or impossible to analyze. Such experiments are fragile and clunky. Extraordinary control measures must be put in place to keep the experimental data pure.

Now compare this with magic shows. Magic tricks test many of the same cognitive processes we study, but they are incredibly robust. It doesn't matter in the slightest that the entire audience knows it is being tricked; it falls for each trick every time it is performed, show after show, night after night, generation after generation. We thought, if only we could be that deft and clever in the lab! If only we were half so skilled at manipulating attention and awareness, what advances we could make!

The idea rapidly took shape: we would bring scientists and magicians together so scientists could learn the magicians' techniques and harness their powers.

But there was just one problem: we were clueless about magic. We didn't know any magicians. Neither of us had ever even seen a real magic show. Fortunately, our colleague Daniel Dennett got us our big break. Dennett is a fellow scientist and philosopher who also happens to be a good friend of James the Amaz!ng Randi, a famous magician and skeptic who has spent decades debunking claims of the paranormal. Randi wrote back, enthusiastically endorsing our idea. He told us that he knew three more magicians who would be perfect for our purposes: Teller (from the magic duo Penn & Teller), Mac King, and Johnny Thompson. All of them lived in Las Vegas and all were personally interested in cognitive science. Apollo Robbins, "the Gentleman Thief," a friend of Teller, joined our group a few months later. Much of this book is based on our interactions with these talented magicians.

Thus began our journey of discovery about the neural underpinnings of magic. We have spent the last few years traveling the world, meeting magicians, learning tricks, and inventing the science of neuromagic. We developed our own magic show and decided to audition at the world's most prestigious magic club, the Magic Castle in Hollywood, California, as bona fide magicians. (For how we did, see chapter 11.)

Magic tricks work because humans have a hardwired process of attention and awareness that is hackable. By understanding how magicians hack our brains, we can better understand how the same cognitive tricks are at work in advertising strategy, business negotiations, and all varieties of interpersonal relations. When we understand how magic works in the mind of the spectator, we will have unveiled the neural bases of consciousness itself.*

So pull up a seat, because *Sleights of Mind* is the story of the greatest magic show on earth: the one that is happening right now in your brain.

*Readers can find relevant citations of the original research studies discussed throughout this book in each chapter's note section.

1

THE WOMAN IN THE CHAMELEON DRESS

Visual Illusions and Magic

Johnny Thompson, the Polish magician, known as the Wizard from Warsaw with a routine of countless corny jokes—"Since I'm part Polish, Irish, and Sicilian, I could have been a drunken janitor who doubles as a hit man"—sweeps onstage in his immaculate tuxedo. Renowned as the Great Tomsoni—"you can call me Great"—Johnny has the affable air of a master conjurer who is about to lead you up (or is it down?) an M. C. Escher staircase of trickery. He has a strong chin, a prominent nose, huge ears, and one of the most wondrous combovers in the world of showbiz.

Imagine for a moment that you are in the audience. The lights dim and Johnny flings his arm toward a bright spotlight enveloping his beautiful assistant, who is clad in a tiny white dress. The Great Tomsoni announces that he will magically change her dress from white to red.

As your eyes focus on the woman, her image is burned deeply into your retinas and brain. Johnny claps his hands. The spotlight dims ever so briefly and then flares up in a dazzling blaze of red light. The woman is suddenly awash in red.

Wait a minute! Switching the color of an ordinary spotlight is not exactly mind-blowing magic. Johnny stands at the side of the stage, looking pleased with his little joke. Yes, he admits, it was a cheap trick, his favorite kind, he explains. But you have to agree, he did turn her dress red—along with the rest of her. Please, indulge him and direct your attention once more to his gorgeous assistant as he switches the lights back on for the next trick.

Johnny claps his hands. The lights dim again. You're wondering why you bought tickets to such a lame magic show when suddenly the stage explodes in a supernova of whiteness. And what do you see? Inexplicably, this time the woman's dress really has turned red. Bright crimson red. She does a couple of turns so you can observe the magical transformation.

The Great Tomsoni has done it again.

Johnny has just created a spectacular illusion based on fundamental properties of your brain's visual system. Visual illusions—which we study for a living—are a particularly palpable demonstration of the systematic illusion spinning that is happening all the time in your brain, at all levels of perception, awareness, and thought. By definition, visual illusions are subjective visual perceptions that do not match the reality of the world all around you.

When you experience a visual illusion, you may see something that is not there, fail to see something that is there, or see something different from what is there. Your perceptions contradict the physical properties of what you are looking at. You can immediately appreciate why visual illusions are useful to magicians. And for scientists, they are indispensable tools for explaining the neural circuits and computations by which your brain constructs its everyday experiences.

The spooky truth is that your brain constructs reality, visual and otherwise. What you see, hear, feel, and think is based on what you expect to see, hear, feel, and think. In turn, your expectations are based on all your prior experiences and memories. What you see in the here and now is what proved useful to you in the past. You know that shadows fall a certain way depending on time of day, that faces are normally viewed in an upright position, and that gravity exerts a predictable influence on all things. When these predictions are violated, your brain may take more time to process the data, or you may focus your attention on the violation. But when everything sails smoothly along, with no surprises, your visual system will miss much of what is going on around you. This is how you drive home without remembering what happened between your office and the driveway.

A fundamental theme of this book is that the brain mechanisms that elicit perceived illusions, automatic reactions, and even consciousness itself essentially define who you are. They evolved along with your bipedal gait and hairless monkey physique. They are the products of an evolutionary path that made it possible for your ancestors to make it through numerous bottlenecks of human history, survive the ice age, and go on to invent agriculture, language, writing, and ever more sophisticated tools.*

You are the result of this epic journey, the likes of which the world has never seen before. Without these innate sensory, motor, and cognitive skills you could not download apps on your smart phone, drive a car, negotiate the interpersonal relationships required to graduate from high school, or even hit a baseball. The reason you can do these things is that, essentially, you are a prediction machine, and you effortlessly and correctly predict almost every event that is about to occur in your life.

Magicians understand at a deeply intuitive level that you alone create your experience of reality, and, like Johnny, they exploit the fact that your brain does a staggering amount of outright confabulation in order to construct the mental simulation of reality known as "consciousness." This is not to say that objective reality isn't "out there" in a very real sense. But all you get to experience is a simulation. The fact that consciousness feels like a solid, robust, fact-rich transcript of reality is just one of the illusions your brain creates for itself. Think about it. The same neural machinery that interprets actual sensory inputs is also responsible for your dreams, delusions, and

*While it is tempting to conclude that we humans have special cognitive skills that other species lack entirely, every time scientists decide that some attribute or capacity distinguishes us from the rest of the animal kingdom, other researchers quickly disprove them. Knacks such as language, tool use, fashion and culture, even dancing are not exclusive to humans but were all considered at one point as defining of and restricted to the human realm.

failings of memory. The real and the imagined share a physical source in your brain.

In coming chapters we will argue, and hopefully convince you, that a surprising proportion of your perceptions are fundamentally illusory. You think you see curvy lines but, when you measure them with a yardstick, the lines are straight. You think you are paying attention, but the pickpocket deftly removes your watch in front of your face. You believe you are aware of your surroundings, but at any given moment you're blocking out 95 percent of all that is happening. Magicians use these various perceptual pitfalls and brain processes against you in a form of mental jujitsu. The samurai invented jujitsu as a way to continue fighting if their swords broke in battle. Striking an armored opponent would be futile, so jujitsu is based on the principle of using an attacker's own energy against him rather than opposing it. Magicians have a similar MO. Their arts are founded on the principle of using your mind's own intrinsic properties against you. They reveal your brain for the liar that it is.

For the red dress trick, Johnny is hacking into your visual system. Comprised of eye and brain, this system should not be compared to an expensive video camera that takes pixel-rich images of the world. Rather, it is a highly evolved kludge of circuits that relies on approximations, guesses, predictions, and other shortcuts to literally construct what might be happening in the world at any given moment.

So, what do we know about those circuits? Exactly what aspects of the brain give rise to visual illusions? How can we probe the visual system to understand the ultimate source of illusions? Full disclosure: often we cannot. Throughout this book, we will be making a distinction between psychological principles and their neural correlates. Take, for instance, post-traumatic stress disorder, or PTSD. The psychological principle that too much stress can lead to PTSD is well documented. But that does not tell you anything about the brain mechanisms involved. To get at the neural correlates of PTSD, you need a neuroscientist to delve into the brain to ferret out the details of what's going on physically inside its circuits.

As for visual illusions, a psychological principle refers to an illusion that occurs when physical reality does not match perception. If your eyes see depth when you look at a painting on a canvas, it must be due to how edges and contours in the image interact in your mind. But that does not tell you anything about how the brain produces the illusion. A psychological principle treats the brain as a black box. It is a "dry" description of perceptions and their presumed underpinnings. A neural correlate is a direct measure of brain activity and anatomy, and it tells you what parts of the brain are used to process the percept, which circuits within those brain areas give rise to an illusion, or even minute details such as what neurotransmitters are involved. It is "wet" biology. We know more about psychological principles than we do about neural correlates, but the gap is beginning to close. Arguably the most exciting breakthroughs in science today are happening in the field of neuroscience.

To understand what neuroscientists are up against when explaining visual illusions, you need to know some nuts and bolts of how your visual system is put together. Your eyes tell you only part of what you are able to "see." The rest is done by your brain in a labyrinth of stages.

The first layer of your visual system consists of *photoreceptors* in your eyes that convert light into electrochemical signals. It is also in this layer where a cardinal attribute of your brain originates: the ability to detect contrast. This property forms the basis of all cognition, including your capacity to see, hear, feel, think, and pay attention. Without it, the world would have no boundaries and your brain could make no sense of itself or anything outside itself.

Naturally, magicians have stumbled onto methods that capitalize on contrast detection, including a stunning illusion called Black Art, which we'll describe later in this chapter.

Information from your retina is funneled into a bundle of fibers called the *optic nerve*, which carries electrochemical patterns into your brain. Everything you perceive enters your brain as patterns. You don't really "see" anything; rather, you process patterns related to objects, people, scenes, and events to build up representations of the world. This information makes a brief stop in the center of your brain, the *thalamus*, before ascending to your *primary visual cortex*—the forebrain's first visual area, and the first of thirty or so cortical regions that, in hierarchical fashion, extract more detailed information about the visual world. This is where you first detect the different orientations of lines, edges, and corners in a visual scene.

Moving up the hierarchy, you have *neurons* that fire in response to contours, curves, motion, colors, and even specific features such as hands and faces. You have neurons that are binocular—they respond to stimulation from both eyes as opposed to one eye alone. Some fire when a target moves left to right; others fire only when a target moves right to left. Still others respond only to up-down or down-up movement. Some respond best to moving edges, or to moving edges of a particular orientation. Thus you go from detecting points of light in photoreceptors to detecting the presence of contrast, edges, and corners, to building entire objects, including an awareness of their color, size, distance, and relation to other objects.

In this process, your visual system makes inferences and guesses from the get-go. You perceive a three-dimensional world despite the fact that a simple two-dimensional image falls on each retina. Your visual circuits amplify, suppress, converge, and diverge visual information. You perceive what you see as something different from reality. Perception means resolving ambiguity. You reach the most plausible interpretation of retinal input by integrating local cues. Consider the full moon rising on the horizon. It looks massive. But hours later, when the moon is high overhead, and is in fact closer to

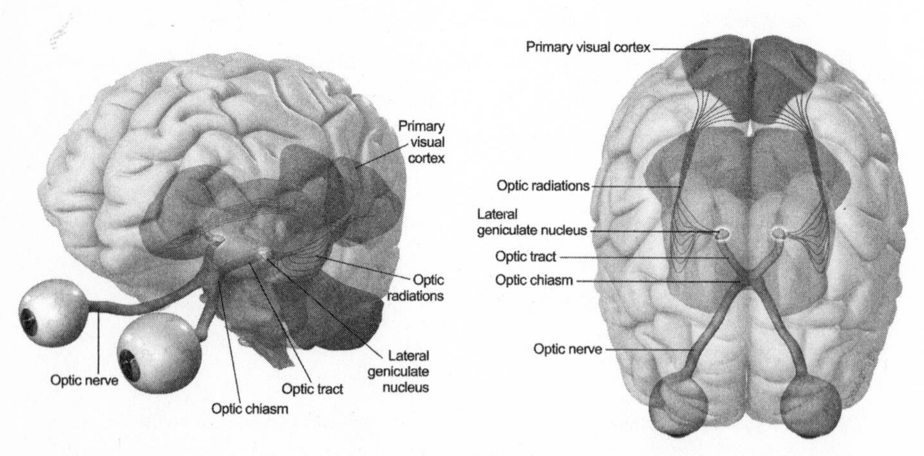

The neurons of the early visual system reside inside the eyes, in the lateral geniculate nucleus (center of the brain) and the primary visual cortex (back of the brain). The wires that connect these brain areas into the visual pathway are in the optic nerve, optic chiasm, optic tract, and optic radiations. (Courtesy of the Barrow Neurological Institute)

you by one-half the diameter of the earth, it looks much smaller. What could explain this? The disc that falls on your retina is not smaller for the overhead moon than for the rising moon. So why does the overhead moon seem smaller? One answer is that you inferred the larger size of the rising moon because you see it next to trees, hills, or other objects on the horizon. Your brain literally enlarges it based on context. This is also why a gray piece of paper can appear dark if surrounded by white or the same sheet can appear bright if it is surrounded by black.

Alas, you simply cannot trust your eyes.

You also make up a lot of what you see. You "fill in" parts of visual scenes that your brain cannot process. You have to do this because of the sheer limitations in the numbers of neurons and neuronal connections underlying your sensory and mental processes. For example, your optic nerve contains all the fibers that send visual information to your brain. Each optic nerve is made up of about a million neural wires connecting each retina to your brain. The individual wires are called *axons*, and each represents one "pixel" of your visual image. Each eye is thus roughly equivalent to a one-megapixel camera. Sounds like a lot, but consider that even your cell phone camera probably has better resolution than that. So how can it be that you have such a rich and detailed perception of the world, when in fact your visual system's resolution is equivalent to a cheap digital camera? The short answer is that the richness of your visual experience is an illusion created by the filling-in processes of your brain.

VISIBILITY AND LIGHT

You might think that visibility should merely require that light fall on your retina. But it is more complicated than that. Not all of the light used by your brain is visible to you. For instance, like all humans, you are bad at accurately estimating the physical light level of your surrounding environment. You don't consciously know how big your pupil is at any given time. Part of the reason for this is that the irises adjust for light level and help to make differently lit environments accessible for neural processing. In low light, your irises open to allow in more photons, and in high light your irises close to keep your retina from becoming blinded by glare. That's why a light level expert such as a photographer must use an objective light level measuring device called a photometer, rather than her own subjective visual estimates of light level, before she can determine the best f-stop to use with her camera lens. But this seems almost like circular reasoning. How can it be that we are unable to accurately quantify the amount of light coming into our eyes due to the change in our irises, yet it must be the brain that controls our irises to optimize the photon density reaching the retina? The answer is that the neural control of the iris does indeed accurately estimate changes in light level, but it does so with circuits that are not connected to the visual circuits that result in conscious awareness. Thus you are only conscious of certain aspects of the scene, such as the relative luminance of objects in the scene, whereas other bits of visual information, such as a quantified measure of overall light level, are handled unconsciously.

Magicians are constantly exploiting these features of your visual system in their tricks. They use illusions of depth in card tricks. They use context to mislead your perceptions. They count on your filling in the missing pieces of a scene. They draw on edge-detecting neurons to convince you they can bend spoons. And they can even draw on specific properties of your visual system to make you momentarily blind—which gets us back to Johnny.

SPOILER ALERTS

Some magicians believe that the secrets behind tricks and illusions must never be revealed. But most agree that some exposure of magic is necessary for the art to thrive, as long as the secrets are revealed carefully, and only to those people who need to know. Jack Delvin, president of the Magic Circle, a leading international society of magic and illusion, puts it like this: "The door to magic is closed, but it's not locked." That is, there are no real secrets in magic; it's all there for everybody to discover. But you have to want it enough to seek it. You have to practice like a demon to gain entry to the club, lest you accidentally reveal secrets through poor performance. And it would be unacceptable for somebody to accidentally run across a secret while reading a magazine or overhearing a conversation—or reading a book.

Because it is necessary to reveal some secrets in order to discuss the neuroscience of magic, we have marked each section of the book in which we reveal secrets. The heading is "spoiler alert." If you don't want to know the magical secrets, or to learn how your brain is being hacked by them, you can skip those portions. Or you can join us in exploring why and how you are so easily fooled.

The Great Tomsoni's red dress trick reveals a deep intuitive understanding of neural processes taking place in your brain. Here is how he did it.

As Johnny introduces his assistant, her skintight white dress lures you into assuming that nothing—certainly not another dress—could possibly be hiding under the white one. Of course that reasonable assumption is wrong.

The woman's slinky, seductive body also helps to focus your attention right where Johnny wants it—on her. The more you stare at her, the less likely you are to notice the hidden devices in the floor and the better adapted your retinal neurons become to the brightness of the spotlight shining on her.

All during Johnny's patter after his little "joke," your eyes and brain are undergoing a neural adaptation. When the spotlight is turned off, visual neurons that have become adapted will fire a rebound response known as an *after discharge.* This response causes a ghostly image of the object to linger for a moment.

You see illusory afterimages such as this every day. Think about a camera flash. It goes off and you are left with a temporary bright white spot in your field of vision that fades to dark. For a fleeting instant, the photoreceptors in the portion of your retina that registered the flash "think" that the whole world has suddenly gone bright and white. They adjust to that brightness level instantly. If the flash is bright enough, it may take seconds, sometimes minutes, for your retinas to completely readapt to the true lighting levels.

Adaptation of motion neurons in your brain also explains the waterfall illusion. If you stare at a waterfall for a minute or more and then shift your gaze to the rocks or foliage next to the flowing water, the stationary objects will appear to flow upward. The illusion occurs because the neurons in your brain that detect downward motion have become adapted to the steady stimulation of falling water, making these neurons relatively less active. Neighboring neurons that detect upward motion are not adapted to the motion and, despite having been at rest, are relatively more active. Since your visual system is set up to see contrast—in this case, neurons adapted to downward motion versus unadapted neurons—your brain makes the net conclusion that something is moving upward. Thus, when you look at the stationary rocks, they magically appear to flow upward for a few seconds.

So now do you see why Johnny's trick works? The neurons of your retina that are selective for the color red adapt to the red-lit dress by reducing their activity. Red photoreceptors are more sensitive to this color than are blue or green photoreceptors. Thus the red-sensitive neurons in your visual system are more adapted and will have a bigger after discharge. In the split second after Johnny dims the lights, you perceive a burst of red as an afterimage in the shape of a woman. It lingers in your brain for about a tenth of a second.

During that split second, a trapdoor in the stage opens briefly, and the white dress, held only lightly in place with Velcro and attached to invisible cables leading under the stage, is ripped from her body. Then the lights come back up to reveal a genuine red dress.

Two other factors help to make the trick work. First, the lighting is so bright just before the dress comes off that, when it dims, you are effectively blinded. You cannot see the rapid motions of the cables and the white dress as they disappear underneath the stage. The same temporary blindness can overtake you when you walk from a sunny street into a dimly lit shop. Second, Johnny performs the real trick only after you think it is already over. That gains him an important cognitive advantage: surprise. You are not looking for a trick at the critical moment, and so you slightly relax your scrutiny.

Afterimages linger in all your sensory systems. When you were a child, you may have learned how to create a muscle memory afterimage by pressing the backs of your wrists outward against a doorframe for a count of thirty, after which your arms seemed to levitate. Indeed, sensory afterimages abound in day-to-day life, and insofar as you are consciously aware of them, they are usually only minor, fleeting impressions or annoyances. But to magicians they are gold.

END OF SPOILER ALERT 

As vision scientists, we are constantly amazed by the clever ways magicians finagle your brain's visual circuitry. Recall what we said about your ability to detect contrast: without it, the world would have no boundaries and your brain could make no sense of itself or anything outside itself.

Well, magicians know all about contrast detection. They stumbled onto it more than a hundred years ago with the invention of *black art.* This is not the abracadabra of ancient wizards and witches but a stage method for producing stunning visual illusions that was discovered by accident in 1875 by a German actor and director, Max Auzinger. The story goes that Auzinger was preparing a dungeon scene for a play and, to make it as fearsome as possible, he lined the room with black velvet. At a critical moment, a black Moor was to appear from a dungeon window and recite his lines. But when the actor playing the Moor put his head in the window, no one could see him. The only things visible were two rows of white teeth floating in air below two white eyeballs.

Auzinger immediately grasped the implications of the illusion. By manipulating black sheets against a black background, he could make objects and people appear and disappear onstage. He could build a magic act that no one had ever seen. Soon his show, "The Black Cabinet," starring himself as Ben Ali Bey, was touring the continent to rave reviews.

Today a black art act, Omar Pasha, is just as popular and, equipped with modern materials and lighting techniques, is no doubt more spectacular than shows mounted a century ago. Owned and performed by Michelle and Ernest Ostrowsky with their son, Louis-Olivier, the show features a character who appears with fluorescent props on a jet-black stage bathed in black light. Black light—what scientists call ultraviolet light—vibrates at a shorter wavelength than visible violet light, and it is called "black" because it is invisible. Fluorescence occurs when one wavelength of light is converted into another. Thousands of substances glow or fluoresce under the influence of black light because the invisible light is converted to visible light, making fluorescent substances glow with seemingly unnatural brightness. Vaseline is electric blue. Fluorite shines bright purple, yellow, blue, pink, or green. Other materials glow red or orange, depending on the chemicals in them.

In the summer of 2009, we saw Omar Pasha in a live performance.* Here's how it goes. As the curtain rises, a man wearing a white turban decorated with red brocade, a white silk brocade tunic, silk pantaloons, a red sash and red cape, white gloves, and red shoes with those little red curlicues on the toes, like one of Santa's elves, bows low. He does not smile—ever. It is Omar Pasha (Ernest Ostrowsky), who resembles a cross between the seductive French president Nicolas Sarkozy and the swashbuckling actor Errol Flynn. Ravel's *Bolero* throbs in the background. The stage floor, side walls, curtains—all that you can see—are pitch-black except for Omar Pasha, who is bathed in black light.

For his first trick, Omar removes a large felt pen from his turban and, with a few broad strokes, draws what looks like a five-foot-high music stand with a gold crossbar on top. The object seems to pop out of nowhere. Then he draws three red candles rising from the crossbar.

*You can view this performance at http://www.sleightsofmind.com/media/blackart. They did not use black lighting in this video, but instead carefully calibrated the lighting so that the camera could not see secret objects when covered by black.

Now it's a candelabra—only it's not a drawing. It's a real three-dimensional object. Omar picks it up and bows slightly, inviting applause. Next, he lights one candle and holds it at arm's length from his right side. He holds a second candle at arm's length to his left. He glances at the flame in a beckoning way. The flame then floats up over his head and descends to the other candle, lighting it. Omar, looking pleased, nods in the direction of the third candle, which is still on the stand. The solo flame takes off again, arcing high over his head, and lands on the third candlestick. After taking a bow, he presses the three candles between his palms and they disappear. The candelabra floats across the stage and parks itself on a table.

For his second trick, Omar picks up a white silk sheet lying crumpled on the floor and snaps it with a flourish. A chair appears out of thin air. Omar walks behind the chair and covers it with the sheet, showing the profile of the empty chair. He then picks up the sheet, filling it with air like a spinnaker, and when it falls, you see that a turbaned young man is now seated in the chair. Omar blindfolds him and grabs a saber. Stepping in front of the man, Omar makes a slashing motion across the man's neck. When he steps to the side, the man is headless. Omar holds the severed head for a few seconds and then places it in the outstretched hand of the headless man. It sure looks real. Then Omar steps in front of the man a second time, and when he moves aside the head is restored. Oh, good, no harm done.

For his third trick, Omar reaches out into space, and a rolled-up poster appears in his hand. He unrolls it, revealing the drawing of a beautiful young woman. He then hangs the poster in midair and unrolls it to reveal the woman herself, who steps through the frame and onto the stage.

Omar covers the woman and the young man in sheets. He makes some magic moves with his hands and the sheets grow and shrink in a ghostly manner. The short sheet with the woman in it grows to man size, and the tall man-sized sheet shrinks to woman size. You know what happens next. The transportation effect is the cleanest version of this particular trick we have ever seen.

Now for the finale. Omar again covers the young man in a silky sheet and invites him to take a few steps forward. Standing behind him, Omar waves his hands, grabs the sheet, and rips it off. The man has vanished. Next, he sets a hula hoop on the ground and steps through it to show it is unencumbered. The young woman steps into it. Omar pulls up the hoop and we actually see her disappear into the ether as the hoop rises and obliterates her body. Finally, he picks up a sheet from the floor and covers himself. As the *Bolero* reaches its crescendo, he disappears from beneath the sheet while it is twice snapped by an invisible hand. The six-minute show is over.

All of these amazing effects are rooted in contrast detection. Your eyes cannot detect anything without some sort of change being present. One way to explain this is through a familiar experience—gazing up at the night sky filled with stars. Imagine you're lying on your back on a warm summer evening under a moonless firmament. All of those points of light are so far away that the area each star activates on your retina is smaller than the area of a single photoreceptor. This means that from your brain's perspective, a star is the smallest thing you can see.

Now imagine you are looking up at the blue sky on a clear day. All those stars are still there, shining brightly, but you cannot see them. During the day, you are star blind. The reason has to do with the comparative amounts of light reaching your eyes during the day and night. At night, a typical star produces 10 percent more light than the surrounding scattered light from the atmosphere. It's a tiny amount but enough to enable your visual system to discern the star. This contrast between foreground and background is the fundamental signal used by your brain to create your mind's image of the star. Without that contrast, your brain's neurons would have nothing to talk about with each other. During the daytime, the blue sky is 10 million times brighter than the blackest night sky. The star that was perfectly visible at night cannot be detected by your visual system because the surrounding sky is so bright that the star's minute contribution of brightness simply cannot be detected as contrast. In the words of Henry Wadsworth Longfellow, "the sky is filled with stars invisible by day."

SPOILER ALERT! THE FOLLOWING SECTION DESCRIBES MAGIC SECRETS AND THEIR BRAIN MECHANISMS!

Omar Pasha created his gorgeous illusions by covering the stage completely with black velvet. When the show began, various objects onstage—the music stand, candles, chair—were also draped in black velvet. With no contrast for our visual system to work with, the objects were invisible to us. He also used black light and fluorescent paints to further decrease the visibility of the black background against the glowing objects onstage.

The entire act is mute. If Omar spoke, his teeth would glow a ghoulish purple. If he did not wear gloves, his fingernails would fluoresce. His eyes shine mysteriously. As Omar prances around the stage, he removes one black covering after another, rendering objects visible. When he replaces the covers, they become invisible. Assistants wrapped in black velvet easily move onto and off the stage without our catching a glimpse of them. Hands in black velvet gloves make flames float through the air. The head is chopped off with the assistance of a black velvet hood. The woman disappears from within a hoop attached to black sheets.

It's not that the black velvet is invisible. Had Omar stuck his white-gloved hand behind the candlestick before it had been revealed, you would have seen a blackened candlestick silhouetted against the glove. No, the trick lies in the contrast, or lack thereof, between the black cloth covering the various objects and the black background of the set and stage.

Magicians are not the only ones who manipulate contrast to make things invisible. Animals do it all the time. It's called camouflage.

Every animal who ever used camouflage is decreasing its contrast as compared to the background, making itself as invisible as possible. With stars in the night sky and with Omar Pasha, reducing contrast means reducing the amount of light against a black background. But another way to decrease contrast is to make yourself the same color, texture, or brightness as the background—like a chameleon, or a stick bug, or a soldier in camouflage fatigues. Contrast is the difference between one object and its surroundings. If there's no difference in color, luminance, or texture, there's no visible contrast, no matter how much light is on the subject.

END OF SPOILER ALERT

Jamy Ian Swiss—with his neatly trimmed Vandyke, combed-back hair, and diamond stud in his left ear—is a magician's magician and czar of close-up card tricks. *New Yorker* writer Adam Gopnik calls Jamy the Yo-Yo Ma of mentalism. Penn and Teller call him "James Bond with a deck of cards for a pistol." Jamy refers to himself as "an honest liar." Being a magician, he says, is the most honest living he has ever made—he promises to deceive you, and then he does.

Boy, does he ever. He lays out four cards facedown, rolls up his sleeves, and waves his elegant hands over the cards, as if stirring up magical currents. His fingers snap and one of the cards is inexplicably faceup—the ace of spades. Another snap. A second card has mysteriously turned over—the ace of hearts. Snap. Snap. Two more aces are faceup. Your brain snaps, too. How could he possibly do that?

Jamy pulls out a deck of cards and shows you a card on top—say, the three of diamonds. "Did you ever see anybody wave their hand over a deck of cards and have it change into another card?" he says as he makes the motion. The three of diamonds turns into the jack of clubs. "You just wave your hand over it," he says, repeating the movement, "and then it changes"—the jack of clubs turns into the six of hearts—"just like that." He continues, "Sometimes you just give the cards a little snap . . ." and he turns card after card into a different card. You cannot see him do anything remotely suspicious—cards seem to float through solid matter under the spell of his agile fingers.

We wanted to meet Jamy because he is one of the best close-up magicians in the business. Many magic tricks involve the use of props (the classic smoke and mirrors) and other elaborate production values. But to master close-up tricks, a magician needs to confound the human visual system. We wondered how deeply a great magician like Jamy has thought about this. Was he intuiting the science? Was he curious about what we know about the brain's inner workings? Many of the magicians we have spoken to have considered these questions, though they haven't had the scientific expertise to isolate the answers.

Of course, that hasn't stopped them from speculating and forming opinions.

Our first meeting with Jamy is in the Three Flags Café at the Marriott Hotel in Monterey, California, four blocks from Fisherman's Wharf and Cannery Row. It is late morning and the place is almost empty. The air smells of coffee and low tide.

We ask to see his Retention of Vision Vanish—a sleight involving the manipulation of a single coin. It was popularized over a century ago by Nelson Downs, a magician from the late Victorian and Edwardian eras known as the "King of Koins." Downs claimed he could palm—that is, hide in one hand—up to sixty coins at a time.

Our friend Eric Mead, the terrific mentalist and magician, had told us that Jamy performs the finest Retention of Vision Vanish he had ever seen. Jamy does not disappoint. With a sly smile, he opens his left hand with a flourish. The palm is faceup, tilted slightly in our direction. He points to it with his right index finger. Then he produces, in his right hand, a shiny fifty-cent piece, pinched between his thumb and first two fingers. His gaze follows the coin as he places it in the palm of his left hand.

Jamy's left hand curls into a fist around the coin, one finger at a time, starting with his left index finger and moving sequentially to his left pinky, like a wave on the Banzai pipeline on the North Shore of Oahu. As his fingers close, we can see the coin as it disappears behind the wave. At the same time Jamy's right hand moves away.

And then it is over. We watch intently as Jamy opens his fist once again and the coin—which we most definitely saw nestled inside his palm—is now gone. Incredible!

SPOILER ALERT! THE FOLLOWING SECTION DESCRIBES MAGIC SECRETS AND THEIR BRAIN MECHANISMS!

Jamy tells us that the retention of vision effect works best with a shiny object. A coin is perfect because he can rotate it as he deposits it into his left hand. This ensures that every viewer sees a flash of light reflected from the lights in the room. That flash creates a brief afterimage, not unlike a flashbulb from a camera, but less intense. You literally see the image of the coin vanish, or fade to nothing, before your very eyes.

Jamy's trick is similar to Johnny's red dress deception in that both exploit afterimages. The difference is a matter of scale, timing, and the specific populations of neurons being adapted. Johnny makes your visual system adapt to a selective target, the red dress. Jamy uses a flash to adapt just the small portion of your retina viewing the coin. He closes his hand over the coin just as its afterimage is created. This buys him a few fractions of a second to remove the coin and hide it in his right hand—while the audience thinks it is obviously in the left. We see it clear as day. The afterimage begins to fade as Jamy's fingers curl to make a fist. But we are already fooled.

Jamy goes on to tell us that it isn't just what he does with his hands that makes the trick work. He uses his whole body. He exaggerates by deliberately shifting his posture to indicate his intentions. Magicians use tension and relaxation to manipulate your judgment of where the hidden object is and is not. (We realized we would have to apply this principle to our audition at the Magic Castle.)

Jamy demonstrates a fake transfer. He will pretend to pass a coin from his right hand to his left. He makes the motion of passing the coin and then accentuates the consequences. His left hand, which presumably received the coin, suggests tension. His right hand, which presumably relinquished the coin, is relaxed, as if it holds nothing. Then Jamy throws his whole body into the act. As he makes the fake pass, he shifts his weight from right to left, as if that side of his body now bears the weight of the coin. He turns his waist, swiveling and slumping his shoulders ever so slightly to the left, as if transferring weight from one hand to the other. He swivels his head as his eyes follow the coin from one hand to the other.

It doesn't matter that the coin weighs less than a sip of coffee. We don't need to shift our bodies with every little slurp from our cups. But Jamy ever so slightly exaggerates every aspect of his fake transfer to convince us otherwise. By combining his adroit act with the coin's afterimage, and by transferring his attention, Jamy loads an incredible amount of suggestive power into the one tiny fake event. He creates an attentional smorgasbord of cognitive clues for us to discover on our own. His technique is so powerful that even though we ask him to do it over and over (a no-no for both spectator and magician), and even though he acquiesces (for the love of science, he explains), we cannot keep ourselves from being deceived again and again. Even though we will ourselves to focus on the relaxed parts of Jamy's body, we cannot help focusing on the tense parts. "That's where the action is," our brain keeps telling us, despite our knowledge to the contrary.

END OF SPOILER ALERT

Jamy has added a cognitive illusion to a visual illusion, a strategy that many great magicians use to convince you that impossible things can happen. Cognitive illusions, which we'll explore in the next chapters, involve higher-level brain functions such as attention and expectations. But before we go there, let's discuss a few more visual illusions that derive from higher levels of the visual hierarchy.

2

THE SECRET OF THE BENDING SPOON

Why Magicians Watch Their Angles

Six weeks after visiting Jamy, we're sitting on the patio of the Cheuvront restaurant and wine bar on Central Avenue in Phoenix, long-stemmed glasses in hand. The light rail train running up the middle of the avenue scrapes and clanks its way northward and makes a stop a few hundred yards away. A sole figure descends onto the platform and walks toward us, a black bag swinging with each step. It's Anthony Barnhart, or Magic Tony to his fans, and he's carrying the tools of his trade—playing cards, a little bag of coins, red sponge balls, and prepared ropes for tricks.

Magic Tony is our mentor and magic instructor, and we're meeting him for another rollicking session of "teach the scientists how to prestidigitate or at least pull off some classic magic tricks without embarrassing themselves when they try out at the Magic Castle." Tony is a big guy with a black crew cut and a jovial demeanor. During the week, he's a PhD candidate in psychology at Arizona State University in Tempe. But on Friday nights he dons his dorky red fish tie ("I just wear it for the halibut") and his leopard print shoes ("It took two leopards to make these, but it's okay because they were babies") and strolls from table to table doing tricks at the Dragonfly Café in North Scottsdale. Customers love him. We do, too.

Tony grew up in Milledgeville, Illinois, where he had a swim coach who taught magic on the side. Along with the Australian crawl, seven-year-old Tony took beginner's magic lessons and was hooked. He also learned a critically important lesson: magic is about entertaining the audience. Young magicians should not focus on methodology at the expense of theatrics.

A magic shop, the Magic Manor, was located an hour away in a strip mall in the neighboring town of Rockford. Like many young boys who fall in love with magic, Tony spent countless afternoons at the shop rummaging through bins and taking group lessons, where he was always the youngest person in the class (and the quickest learner). He attended Tannen's Magic Camp on Long Island for two years in a row. His favorite memory of getting fried (that is, fooled badly) happened in his dorm room in the middle of the night. He was rooming with two other campers. At about two in the morning, the counselor woke them up and said to Tony, "Think of a card." He did so, and the counselor subsequently named the card that Tony had sleepily thought of (it was the seven of clubs). "I'm still not sure how he did it," says Tony. "He must have primed me in some way, but I can't be sure. I sort of like not knowing."

Today Tony is going to teach us two methods used in the Ambitious Card routine. This famous trick can be done in an infinite number of ways, but the ones we are about to learn are especially germane to how magicians trip up your visual system. The magician asks you to choose a card, any card, from a deck. You do so and then place the card in the middle of the deck. The magician snaps his fingers over the deck and voilà—your card has mysteriously risen to the top. It is an ambitious card—it rises through all the other cards every time.

The routine is renowned in the annals of magic as the trick that fooled Harry Houdini. In the early decades of the twentieth century, Houdini was the most famous magician in the world. Whereas he had earned supreme confidence in his abilities to pull off spectacular escapes, he was perhaps too confident in his abilities in close-up magic. With fulsome bravado, Houdini issued a challenge to all magicians: Show me any trick three times in a row and I'll tell you how you did it.

At the Great Northern Hotel in Chicago in 1922, a gifted magician, Dai Vernon, met the challenge by demonstrating his version of the ambitious card routine. Vernon, known as the "Professor," was more than a match for Houdini. He was one of the best sleight of hand artists who ever lived and, with another magician, Ed Marlo, possibly the most influential card magician of the twentieth century. Vernon was a brilliant inventor of close-up effects with cards, coins, balls, and other small items.

Vernon asked Houdini to choose a card and sign it, in ink, with his initials. The card went into the middle of the deck. Vernon snapped his fingers. Houdini's card was on top.

Houdini was stumped. "You must have a duplicate card."

"With your initials, Harry?" asked Vernon.

Vernon repeated the trick three times, using a different method each time. Houdini was incensed. He couldn't figure out how it was done. Vernon did the trick four more times. Still Houdini was fooled—though he never admitted it publicly.

Sleight of hand magic, when done well, is miraculous to behold. (The word "sleight" comes from Old Norse and means cleverness, cunning, slyness.) It is usually performed close-up, within a few feet of a spectator. There are hundreds of different sleights. Some involve misdirecting your attention (we'll get to those in chapter 4). Others exploit foibles of your visual system. Indeed, the role of visual perception in sleight of hand is fundamental to magic.

SPOILER ALERT! THE FOLLOWING SECTION DESCRIBES MAGIC SECRETS AND THEIR BRAIN MECHANISMS!

It's no coincidence that magicians use decks of playing cards to convey their magic. Cards are remarkable in that they are stiff, yet very thin. They fit inside the palm of your hand and can be hidden easily. They can be shuffled, fanned, flipped, palmed, cut, gripped, and pocketed. Our first lesson today is the *double lift*, probably the most basic and most central sleight in the magician's repertoire—and a key feature of Ambitious Card routines. The trick is to turn over two cards on the top of the deck while making it look as if you are flipping only one. It's that simple. But when it is used at the right time in concert with other types of misdirection, it is utterly astonishing. Dai Vernon was a master of the double lift. Say your card is the ace of clubs. The magician fans the cards and you put the ace in the deck. As he closes the fan, he puts one card on top of the ace and surreptitiously marks the spot, called a break, with his pinky finger. He makes a quick cut of the cards so that the ace is now the second card from the top. Then comes the double lift. He lifts two cards so that the ace is faceup, on top. It is the ambitious card.

The magician smiles and says, "Yes, it is an ambitious card." He double flips the cards facedown once again and then takes the top card (which you think is the ace but of course it is not) and puts it in the middle of the deck. He snaps his fingers and turns over the top card, which is—the ace! It is surely ambitious, and you are dumbfounded.

Magicians train for thousands of hours to double-lift without revealing that they are "handling" the cards. They must train their fingers to deftly lift two cards while convincing you that they are lifting only one. This involves various maneuvers such as putting a small crimp in the two cards so that when they are facedown the magician can feel them as one. When the cards are flipped, the crimp is released and the cards lie flat. In mastering this sleight, magicians must be able to make the moves without paying attention to what they are doing. Vernon, in a 1961 book, *Stars of Magic*, warned that many magicians screw up the double lift because they are terrified that the two cards are going to separate. "A playing card," he said, "is a light and delicate object and should not be turned over like a cement block."

So how does the double lift fool you each and every time? Why can't your visual system track the cards correctly? It has to do with your center of vision. If you were going to detect two cards pressed firmly together moving as a unit, you would have to put your eyes inches in front of the magician's hands and stare at the cards as if under a magnifying glass. Even then you might miss the sleight of hand.

The reason for this is that your visual system has very poor resolution except at the very center of your gaze. The cards are so thin that your vision is not up to the task of distinguishing them, especially in the hands of a skilled card sharp. Your center of gaze is called the *macula*—the region near the center of your retinas packed with photoreceptors. It, along with the *fovea* (the very center of the macula and the part with the very highest resolution), is responsible for high-acuity vision. It's a piece of your anatomy that is so specialized it has its own set of diseases, including age-related macular degeneration. In fact, macular degeneration is the most common form of blindness in older people, as maculas slowly die over the course of a few years. Without maculas, you can see only with your peripheral vision, which has very low resolution. You navigate by seeing the world in terms of what appears off to the sides of your head.

END OF SPOILER ALERT

Tony shows us another way to do the Ambitious Card routine called the Vernon Depth Illusion (also known as the Marlo Tilt because the two magicians developed it independently). Long after Houdini died, Vernon continued to refine the trick with diabolical insight into visual processing.

In this sleight of mind—captured on rare film footage in the 1950s—Vernon asks you to choose a card and sign it.* He takes the card and clearly sticks it into the middle of the deck, slowly and purposefully, so there can be no mistake that it's your card. Then he flips the top card of the deck, and voilà!—it's your signed card.

SPOILER ALERT! THE FOLLOWING SECTION DESCRIBES MAGIC SECRETS AND THEIR BRAIN MECHANISMS!

It is an incredible, astonishing, maddening event. Here is how he did it.

After Vernon receives your card, he twists it slightly and sticks it partway into the center of the deck from the back. The twist ensures that the card does not enter the deck. Instead, it forces others cards to protrude where you are looking—at the front of the deck about halfway down. These pushed-out cards reinforce the idea that Vernon

*For a sample of Vernon's deftness, see http://www.sleightsofmind.com/media/vernondepthillusion.

is really planning on pushing your card into the center of the deck. But it's a ruse. While resquaring the deck (pushing the cards back in), as if to fix his mistake, Vernon tilts the back of the top card of the deck slightly upward. From the perspective of where you are standing, you cannot see the tilt, though there is now a gap of almost a centimeter between the top card and the next card down, as seen from the back of the deck.

Vernon then takes your signed card and slips it into the deck at the bottom of the unseen gap. From your vantage point it looks like it is going into the middle of the deck, but in fact it is now the second card down. You don't notice this discrepancy for two reasons. First, from your perspective, you can't see the tilt of the top card. It never occurs to you that he could be sliding your card into the second-card-down position, directly under the top card.

Second, your visual system convinces you that your card is much farther down than the second position. It looks to be in the middle of the deck, in approximately the same position as when Vernon first "accidentally" pushed out the cards with the twisted card. You saw other cards pushed out as it was "inserted." But did you really see it go in?

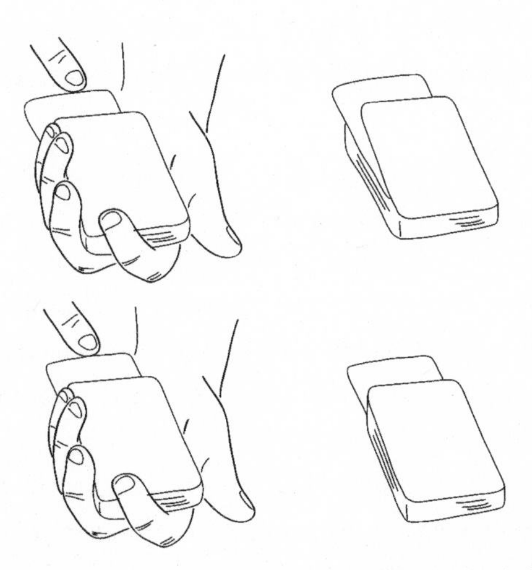

The magician can push the card into the middle of the deck (below right) or just under a tilted top card (above right). Either way, the card looks like it's going into the middle of the deck from the vantage point of the spectator (left column). (Drawn by Jorge Otero-Millan)

Obviously not, but your visual system also tells you that your card is now occluded by the top of the deck. Your angle of perspective tells you that your card is being inserted. And your three-dimensional vision tells you that your card must be in the middle of the deck—about twenty-five cards down from the top card.

Of course this logic is all wrong when the back of the top card is tilted up during the second insertion attempt. Afterward, a very innocent motion of Vernon's hand allows the tilted card to drop, and the gap is now closed. Your signed card is now in perfect position to be revealed by a double lift. Vernon tells you that your ambitious card has risen to the top, and there it is. Then he compounds your sense of awe by saying, "Let me show that to you again." He double-lifts the two faceup cards back down, and then he removes the top card (which is not your signed card, though you think it is) and actually puts it into the middle of deck. And you know the rest. Your signed card is now on top.

Two normal depth perception cues—occlusion and perspective—have conspired to fool you. These processes are automatic and occur without your being aware of them, which is why the trick works. Remember we said your brain constructs reality? In this case, your visual system is telling you what is "real," but it is a hapless victim in the hands of a skilled magician.

Occlusion refers to the fact that if one person is partly hiding behind another person, you naturally assume that the person who is not occluded is closer to you. The same goes for playing cards. This is a logical deduction made by your brain, done automatically and virtually instantaneously, without conscious thought.

Again, Vernon fools your visual system. Because you "see" your card being inserted into the middle of the deck, well then, the other cards must be on top. They are occluding your card, which must be fairly far down the deck.

Nobody knows where occlusion is computed in the brain, but it presumably happens high enough in your visual system that the relevant neurons encode individual shapes. Neurons that become active early in your visual pathway detect only small features of the world—edges, corners, curves. To put together an entire shape and see an object of interest (a person, a card), you need shape-selective neurons that combine the outputs from early feature detectors. Following this logic, you need an even later level of computation that can determine that a neuron's favorite shape is being occluded. In this way, your visual system builds your depth perception like an automobile assembly line, one piece at a time, until you end up with a percept rich in depth.*

Also, Vernon is hacking into your brain's drive to understand the world in perspective. Linear perspective rests on the fact that parallel lines, such as those in a railroad track, appear to converge in the distance (the Leaning Tower illusion in chapter 3 is based on this phenomenon). Your visual system interprets convergence as depth because it assumes parallel lines will remain parallel.

In Vernon's card trick, size perspective comes into play. If two similar objects appear different in size, your visual system assumes the smaller one is more distant. Here, the signed card is slightly smaller on your retina, which means it must be farther away. It must be going into the middle of the deck, based on all the other clues you are seeing.

END OF SPOILER ALERT

In the early 1970s, a new magic superstar swept onto the world stage. His name was Uri Geller—a tall, lanky Israeli with a Beatlesque mop of black hair, dark brows, and a penetrating stare. A charismatic stage performer, Geller could bend spoons, make watches stop or run faster, telepathically read hidden drawings, and otherwise blow people's minds with his "supernatural powers."

It was an era of unalloyed credulity.

Perhaps it was the drugs. When you place a windowpane of LSD on your tongue and watch the world transform into a Salvador Dalí landscape of radiant colors, shimmering geometric shapes, and morphing phantasmagoria while your sense of self dissolves—well, why can't someone bend a spoon with his thoughts?

Perhaps it was Cold War paranoia. The CIA believed that the KGB had learned how to exploit extrasensory perception or remote viewing. Enemy spies could penetrate our secrets from halfway around

*Throughout this book we use the terms "perception" and "percept" as synonyms.

the world using their telepathic powers. They could stop heartbeats from a distance (for an amusing look at his era, see the film *Men Who Stare at Goats*).

Perhaps it was one of those bizarre moments in history when unusually large numbers of otherwise rational people are seduced by magical thinking. New Age fads emblazoned the wonders of tarot cards, I Ching, Kirlian photography, crystal power, dowsing, astrology, and new approaches to personal development in harmony with planetary evolution.

Geller, at the forefront of this craze, was best known for his ability to bend spoons. "Isn't it amazing?" he would marvel while holding a spoon at its neck, stroking it gently but rapidly with his index finger. Slowly, like an acrobat doing a languorous backbend, the spoon would bend, and bend, until it flopped at an askew angle. Spoon magic.

Millions of people were taken in by Geller's act, until famed debunker of paranormal claims, James the Amaz!ng Randi, stepped up to throw a dose of cold water on Geller's hot act.*

Geller once said he performed his feats through supernatural powers given to him by extraterrestrials. Randi came along and said that Geller's feats were parlor tricks. He repeated them all, explaining how each was done—spoon bending, mind reading, stopping watches, dowsing, all of it. "Magical thinking is a slippery slope," Randi says during his demonstrations. "Sometimes it is harmless, other times quite dangerous. I am opposed to fakery. I expose people and their illusions for what they really are."

For example, Randi explains that mentalists have been duplicating hidden drawings for years. A person draws something on a piece of paper and hides it, and the magician reveals what is on the drawing. Sometimes the magician turns his back and covers his eyes while the drawing is made. Randi wonders, "Why cover your eyes with your back turned?" He demonstrates. A small mirror concealed in the palm of his hand as it covers his eyes shows exactly what the person is drawing.

But despite Randi's efforts to expose Geller as an illusionist, people

*You can watch it at http://sleightsofmind.com/media/RandiGeller.

kept believing. Even some scientists were taken in. In 1975 two researchers of paranormal psychology at the Stanford Research Institute, Russell Targ and Harold Puthoff, tested Geller and concluded he had performed successfully enough to warrant further serious study. They called it the "Geller effect." Brain waves, they said, could affect pliable metal.

Danny Hillis, a renowned computer scientist and amateur magician, has an explanation for why scientists are particularly gullible to the Gellers of this world. "The better the scientist, the easier it is to fool them," he says. "Scientists are honest people. They don't know how low magicians will stoop and are not trained in deliberate deception."

For example, Hillis once showed a magic trick to Richard Feynman, the Caltech physicist widely regarded as one of the most brilliant people who ever lived. "I'd do the trick and challenge him to figure it out. He'd go off for a day or two, think it through, and come back with the correct answer," says Hillis. "Then I would repeat the trick using an entirely different method. And it drove him crazy. He never got the meta principle that I changed methods. This may be because of how scientists are trained to use the scientific method. You keep doing experiments until you find the answer. Nature is reliable. The idea that someone would switch methods just flummoxed him."

SPOILER ALERT! THE FOLLOWING SECTION DESCRIBES MAGIC SECRETS AND THEIR BRAIN MECHANISMS!

Spoon bending can be done many ways. Here is how Tony taught us.*

He starts with three spoons and has someone pick one and examine it. He asks that person to put the spoon to his forehead—Tony demonstrates by putting a spoon to his own forehead—and tells the spectator to report when it starts feeling warm.

As Tony brings his spoon down from his little demonstration—while everyone's attention is focused on the poor sucker holding

*Tony learned this method from the mentalist Alain Nu.

a spoon to his forehead—Tony simultaneously bends both of his spoons ninety degrees at the neck.

This is the essence of spoon bending. The spoons are bent before the illusion is created. Magicians call it *ratcheting.* He bends the first one in his right hand with his thumb while holding the stem of the spoon in his fist. He simultaneously bends the second spoon at the neck by pushing the bowl against the inside of his inner right wrist. The maneuver is very clean and natural. It's meant to look as though he is merely bringing the spoons together into his right hand. In any case, everybody's attention is on the guy holding the spoon to his head. Meanwhile, Tony quickly transfers the now bent left-hand spoon into his right hand. He holds the two spoons between his right thumb and forefinger so that the bends of the two spoons touch each other knee to knee. It appears that he is holding two unbent spoons that are crossed at their necks.

Tony then shakes the spoons and "lets them wilt." It looks as if the spoons become soft and floppy and the necks slowly bend. Actually, he is allowing the bent spoons to turn slowly between his fingers so that the bends are in the same direction, and the bowls eventually hang down. While the spoons are bending, Tony takes a brief break and retrieves the third spoon from the spectator with his free hand. He redirects everyone's attention back to the bending spoons by saying that he is concentrating on them. His mind is bending them. Meanwhile, he surreptitiously bends the third spoon against his leg and then holds it so that only the stem is visible.

When the two "wilting" spoons are completely bent, Tony hands

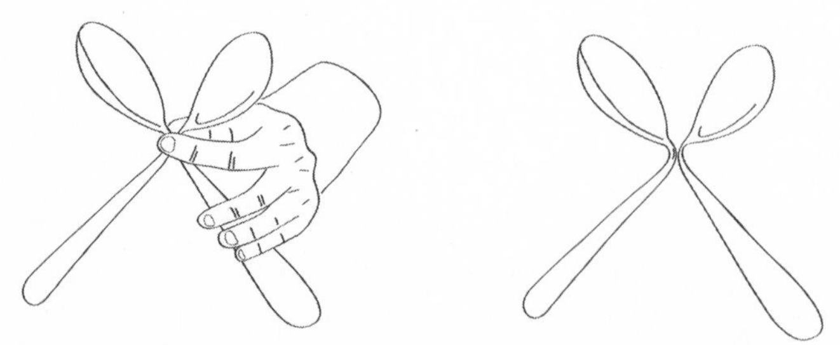

The principle of good continuation helps you see the spoons are crossing when they are held by the magician (left), despite the fact that they are actually bent (right). (Drawn by Jorge Otero-Millan)

them back to the assisting spectator and says, "Now let's try that again." He holds the third spoon in both hands so that the stem is pointed vertically from behind his two interleaved hands. Neither the bowl of the spoon nor the now extant ninety-degree bend in its neck is visible. The audience assumes the spoon is still straight, since the spectator just inspected it.

Tony begins to concentrate on the third spoon, and slowly, excruciatingly, without his applying any perceptible pressure, the stem of the spoon folds until its neck is bent toward him at a ninety-degree angle. Tony hands the bent spoon to the spectator, the audience applauds, and the routine is over.

A few critical psychological concepts help fool you into thinking the spoons must be straight when in fact they are already bent. The first is what visual scientists call *amodal completion*—the process by which an object that is occluded by a second object appears whole to you, even though it's occluded. Imagine you are sitting in for one of our magic lessons here in Phoenix with Magic Tony. You're at Cheuvront, munching on a Spanish cheese plate of manchego and queso de País, glass of Rioja in hand, and looking out across the vast Sonoran desert in between tricks. You notice a jackrabbit. It jumps three bounds and lands partially behind a massive four-armed saguaro cactus, with only its hindquarters sticking out, fuzzy white tail twitching. Does the rabbit still have a head? Of course it does. But how do you know? You can't see it. How is it that your brain informs you about the shape of the hidden part of the rabbit behind the cactus? What if we were not discussing a rabbit but a blank rectangular surface that sticks out from one side of the cactus instead? In that case, you could not know from your experience how big the occluded part is, because rectangles, unlike rabbits, can have any size. But now imagine the rectangle poked out on both sides of the cactus so that you could see all four corners of the rectangle, but the middle remained occluded. Now, despite the fact that most of the surface is occluded, you have a very strong impression of how big the object is and what shape it takes—even though you can't truly know what's going on with the portion of the surface that's behind the cactus.

In the case of the rabbit, your brain has mapped a three-dimensional biological model of a jackrabbit, and it makes perceptual guesses as to what the occluded part of the animal must look like. That's very helpful, especially if you are hunting rabbits. And in the case of the rectangle, your brain can make certain perceptual guesses but not others, depending on how much information it has.

Tony took advantage of amodal completion when he pinched the two bent spoons between his thumb and forefingers. Because the stem from spoon number one lined up with the bowl of spoon number two, and vice versa, each spoon looked straight; amodal completion inappropriately completed both objects behind Tony's fingers. Tony explains that this process obeys the law of "good continuation," first codified by the German Gestalt psychologists of the turn of the century.

WHY GOOD CONTINUATION IS SO GREAT

Good continuation is the process by which your brain makes things seem whole based on sparse information. Amodal completion is one example of good continuation, but there are many others. We already mentioned filling in. The world is too large and too complex for you to see every item in it. When you look at a pebble-strewn beach or intricately woven Persian carpet, your brain is not resolving every pebble or every stitch of fabric. You don't have enough cells in your retina for that. You see a small portion of beach or carpet and fill in the rest. Good continuation is so integral to a plethora of brain mechanisms that Tony thinks it is the most exploited principle in all of magic.

To see how clever your brain is at filling in, try the Ganzfeld procedure. (Ganzfeld is German for "the entire field.") First, cut a Ping-Pong ball in half. Then tune your radio to static. Lie down, tape a half ball over each eye, and wait. Within minutes you will experience a flood of bizarre sensations. Polar bears cavort with elephants. Your long-deceased uncle comes into view. Whatever. Your brain cannot deal with zero sensory input, so it makes up its own reality. The point here is that your brain is constantly making up its own reality whether it receives actual reality-driven input from your senses or not. In the absence of sensory input, your brain's own world making machinations keep on truckin' nevertheless. That's why solitary confinement is considered a punishment in our prison system. You might think that solitary confinement would be a relief from the dangers and unpleasantnesses of prison life. But it is just about the worst thing that you can do to prisoners, because they lose touch with reality. Many consider the practice a form of torture, and volumes have been written on the negative psychological effects of solitary confinement. Prisoners eventually report having hallucinations and other types of psychotic reactions. That is, they begin to believe the illusions.

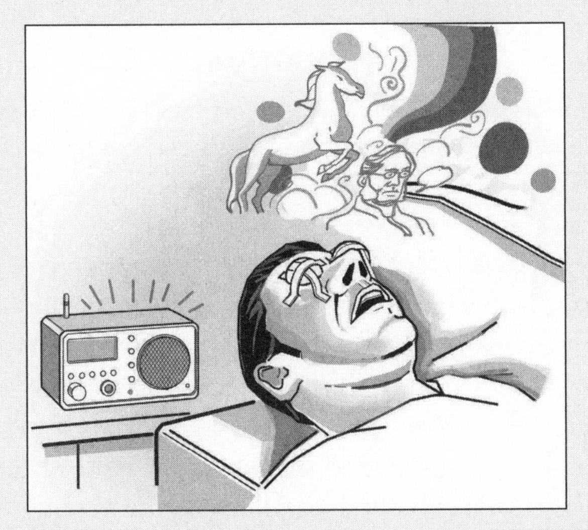

How to hallucinate using Ping-Pong balls and a radio. ("Hack Your Brain," republished with permission of Globe Newspaper Company, Inc., from a 2010 edition of the *Boston Globe* © copyright 2010)

Have you ever wondered how a magician saws a woman in half? The illusion is based on two things—a hollowed-out box and your brain's desire for good continuation. When the woman lies down in the box, you see her head at one end and her feet at the other. Your brain tells you that she is supine and in one piece. Actually she is not lying down flat. The box is constructed so that the head protruding from one end and the feet sticking out the other end belong to two different women. The illusion is often enhanced by a painting of her supine body on the side of the box. How easily you are fooled.

Some mechanisms behind good continuation are becoming well understood. For example, in the visual system, good continuation depends on the orientation and spatial position of lines you may be looking at. When the relative position and orientation of two or more line segments are in alignment, you may see a contour. When two or more lines having similar orientations are positioned close together with their ends aligned, you may notice that individual segments are more visually salient: they pop out against the background. But if the separation between segments, or the differences in their orientations, is too great, good continuation fails and the segments are more difficult to discriminate from the background.

Charles Gilbert and colleagues in his laboratory at the Rockefeller University have found a physical basis for good continuation in the visual system. Recall that neurons in your primary visual cortex are tuned to specific orientations—they prefer, say, horizontal line segments or vertical ones. Such specialized neurons are found in different parts of the primary visual cortex so that your brain can integrate information well beyond the boundaries of single neurons. It turns out that neurons with similar attributes are connected via horizontal fibers that travel long distances in the primary visual cortex. Your mind's eye can "see" the rabbit behind the cactus because of the long-range connections between similar types of neurons in the cortex. The same processes could play a role in more cognitive types of visual perception, which we will discuss more fully in later chapters.

A second concept behind the spoon illusion has been documented. When spoons are shaken just so, they suddenly appear floppy. The illusion occurs because your visual system has two different mechanisms for seeing lines: one that specializes in edges and another that specializes in the ends of lines. To detect the edge of a line, you rely on neurons in your primary visual cortex. To localize the ends of a line, however, you call on *endstopped cells* that are tuned to respond to the ends of long contours.

Some orientation and endstopped neurons respond especially well to moving stimuli, such as the stem of a shaking spoon. But the timing of their responses is different. Your brain perceives the orientation of lines faster than the ends of lines. Thus the stem of a shaking spoon appears to move before the ends move—giving rise to the illusion that the spoon is bending.

END OF SPOILER ALERT

As romantic as it may be to conclude that thoughts can bend spoons or levitate tables or that psychic powers, clairvoyance, and mind over matter are real phenomena, the consequences of such beliefs can be painful, or at least embarrassing. When Susana was about eight years old, she got it in her head that she should be able to walk through barriers by sheer mental effort. The heavy wrought iron gates of her grandparents' apartment building in Santander, Spain, seemed ideal for the experiment. When the adults were down for siesta, she sneaked out and ran down the three flights of stairs leading to the entrance of the building. She was determined and she ran at full speed, headfirst. Surprisingly, the wrought iron didn't budge, as evidenced by the small scar she still has on her left temple. It took more than a decade for her to confess to her family that she hadn't accidentally tripped that day.

Charlatans and frauds abound, taking advantage of unsuspecting or desperate customers who honestly believe in psychic abilities. These customers are inevitably cheated of their money or worse: sometimes they are persuaded to reject proven medical treatments in favor of various sorts of psychic interventions. When a psychic, faith healer, medium, or charlatan appears to defy the laws of nature, there is always an illusion involved. It's our job to discover how they work. And that's part of what this book is about.

3

THE BROTHER WHO FAKED A DOME

Visual Illusions in Art and Science

Vision scientists like us seek to understand how we see, from both a psychological and a biological perspective, and our discipline has a long tradition of studying visual artists such as painters and sculptors. Scientists did not invent the vast majority of visual illusions—painters did. The visual arts often preceded the visual sciences in the discovery of fundamental vision principles, through the application of methodical—although perhaps more intuitive—research techniques.

Likewise, magicians—as the world's premier artists of attention and awareness—have made their own discoveries. This is what drew us to their footlights, card tables, and street performances. We want magicians to help us understand cognitive illusions in the same way that artists have revealed insights about visual illusions. And in fact visual illusions are a bit like magic tricks on the page. In this chapter we'll take a brief tour of some of our favorites.

Artists have been utilizing visual illusions since the fifteenth century, when Renaissance painters invented techniques to trick your brain into thinking that a flat canvas is three-dimensional or that a

In the early decades of the seventeenth century Dutch painters developed still-life easel paintings with trompe l'oeil realism. (*The Attributes of the Painter* by Cornelius N. Gysbrechts. Réunion des Musées Nationaux / Art Resource, N.Y.)

series of brushstrokes in a still life is a bowl of luscious fruit. They figured out linear perspective—the notion that parallel lines can be represented as converging so as to give the illusion of depth and distance. (Again, think of train tracks heading toward the horizon.) They realized they could manipulate atmospheric effects by making tones weaken and colors pale as they recede from view. They used the horizon or eye level as a reference point to judge the size and distance of objects in relation to the viewer. They used shading, occlusion, and vanishing points to make their paintings hyperrealistic.

Trompe l'oeil is a French term that means "trick the eye." It flourished in the seventeenth century in the Netherlands. The lifelike pictures appeared to jump from the frame.*

*An early and perhaps apocryphal example of trompe l'oeil, reported by Pliny the Elder, is the legendary competition between two renowned painters in ancient Greece, Zeuxis and Parrhasios. Each artist brought a covered painting to the contest. When Zeuxis unveiled his work, his painted grapes were so realistic that birds flew from the sky to peck at them. Convinced of his victory, Zeuxis tried to uncover

The "dome" of Saint Ignatius church looks like a real dome from this vantage point. (Flikr.com)

Trompe l'oeil is sometimes used on a large scale to suggest entire parts of buildings that do not actually exist. The architect of the Saint Ignatius church in Rome, Horace Grassi, had planned to build a cupola but died before finishing the church, and the money for the cupola was used for something else. Thirty years later, in 1685, the Jesuit artist Andrea Pozzo was asked to paint a fake dome on the ceiling over the altar. Pozzo was already considered a master in the art of perspective, and yet what he accomplished could hardly be believed. Even today, many visitors to Saint Ignatius's are amazed to find out that the spectacular cupola is not real but an illusion.

Architects soon realized that they, too, could manipulate reality by warping perspective and depth cues to create illusory structures that defied perception. Need a big room in one-fourth the space? No problem. Francesco Borromini accomplished just that at the Palazzo

---

Parrhasios's painting to confirm the superiority of his work. He was defeated, however, because the curtain he tried to pull back was Parrhasios's painting itself.

This hallway is much shorter and the sculpture is much smaller than they appear. (Flikr.com)

Spada, a palace in Rome that we visited a few years ago. Borromini created the illusion of a courtyard gallery 121 feet long in a 26-foot space. There's even a life-size sculpture at the end of the archway. Not really. The sculpture looks life-size but is actually just two feet tall.

Closer to home and to magic is the Grand Canal concourse at the Venetian Hotel and Casino in Las Vegas. The first time you step onto the concourse, you feel a sudden onset of twilight. That's exactly what Susana's mother, Laura, experienced when we first took her to Las Vegas while planning our conference. We descended from our suite after a room service lunch. Stepping out of the elevators and onto the concourse, she said, "Oh, it's gotten so dark outside." Susana asked her what she meant. "The sky," Laura said. "It's gotten dark so early."

"But, Mamá," Susana explained, "we're still inside. You see the black spots in the sky? They are sprinkler heads."

Mouth agape, Laura examined the incredible illusory sky, with its five shades of rococo blue—peacock, azure, cerulean, turquoise, and aquamarine—and wisps of mare's tails, stratus, and cirrocumulus clouds. Laura considered it for a minute before turning to Susana and saying, "Well, why did you tell me so soon? I would have liked to enjoy it a little longer."

Another great illusionist is the Dutch lithographer and woodcut artist Maurits Cornelis (better known as M. C.) Escher. Early in his career, Escher carved realistic scenes based on his observations and travels. Later, he turned to his imagination, rendering some of the most brilliant visual illusions in the history of art. When he was in high school, one of Steve's favorite posters was an Escher print of the never-ending staircase (*Ascending and Descending*, 1960), in which a group of robed monks perpetually climb or descend an impossible staircase situated at the top of a temple. It was impossible because it circled around on itself and never ended. So how could it be drawn if it was physically impossible? Escher must have cheated somewhere in the print and failed to depict the proper structure of a real staircase. But Steve couldn't find it, no matter how closely he looked. He realized he should examine the structure as a whole to see if there was a small systematic warp along the entire structure that allowed for the illusion.

And that's when Steve found that he couldn't look at the structure globally. He could only really see one area of the staircase at a time. His vision could process the details of the staircase when he centered his gaze on a specific part. But when he did that, every other area of the staircase, in his visual periphery, was left in a blur. And he realized that that was how Escher must have done it: since you can see only one local area at any given time, small, gradual errors along the entire structure could not be seen with the naked eye.

This effect challenges our hard-earned perception that the world around us follows certain inviolable rules. It also reveals that our brains construct the feeling of a global percept by sewing together multiple local percepts. As long as the local relation between surfaces and objects follows the rules of nature, our brains don't seem to mind that the global percept is impossible.

Susana's formal introduction to visual illusions came in 1997 when she arrived at Harvard University to study under David Hubel and Margaret Livingstone. At the time, Harvard was the mecca for the study of illusions, and in fact this is where she met Steve. Not only were Livingstone and Hubel leading the field in the study of illusions in the brain, but a number of Harvard psychologists were discovering an array of completely new phenomena.

As part of her postdoctoral training, Susana decided to choose a visual illusion and investigate its effects. Leafing through an art book, she found the perfect playground for her curiosity: op art, a field that explores many aspects of visual perception, such as the relations between geometrical shapes, variations on "impossible" figures that cannot occur in reality, and illusions involving brightness, color, and shape perception.*

Susana settled on op artist Victor Vasarely, whose Nested Squares series exhibited an odd illusion: the corners of the squares looked brighter than their straight-edged sides. But the effect wasn't just about the lightness of the corners, because if Vasarely reversed the order of the nested squares from white-to-black (center to exterior) to black-to-white, now the corners were darker than the sides. So it seemed to be an illusion concerning contrast, and not lightness per se.

Susana searched the vision research literature and found that only a couple of people had discussed this effect previously and nobody had investigated its neural bases. And no one had looked at shapes other than squares. Squares are a special type of shape in which all of the corners are convex (all point away from the center of the square). But nobody had examined the effect for nonsquare shapes with concave corners or for shapes with corner angles other than 90 degrees. Susana realized there were many aspects of this illusion that she could study perceptually, followed by physiological research in the brain.

After several years, first as a trainee at Harvard and later as the director of her own research team, Susana learned one of the most fundamental secrets of the visual system. The previous dogma in the field had been that neurons in the first few stages of the visual system were most sensitive to the edges of object surfaces. Susana's results showed instead that neurons of the visual system are more sensitive to the

*The op art (for "optical art") movement arose simultaneously in Europe and the United States in the 1960s. Unlike the artists that preceded them, op artists did not use illusions merely as a means to achieve a desired perceptual effect such as distance or volume. The illusion itself was the goal.

Vasarely's *Utem* (1981). Nested squares of increasing or decreasing luminance produce illusory diagonals that look brighter or darker than the rest of the squares. (Courtesy of Michèle Vasarely)

corners, curves, and discontinuities in the edges of surfaces, as opposed to the straight edges that had previously been thought to reign.

Op artists were also interested in kinetic or motion illusions. In these eye tricks, stationary patterns give rise to the powerful but subjective perception of illusory motion. An example is *Enigma* by Isia Leviant.

Reinterpretation of *Enigma* (Created by and courtesy of Jorge Otero-Millan, Martinez-Conde Laboratory, Barrow Neurological Institute)

This static image of regular patterns elicits powerful illusory motion in most of us and has generated an enormous amount of interest in the visual sciences since it was created in 1981. However, the origin of the illusion—is it the brain, the eye, or a combination of both?—remains, appropriately, an enigma.

In 2006 we designed an experiment to probe this question. We asked observers to say when illusory motion sped up or slowed down as they looked at the image. At the same time, we recorded their eye movements with high precision. Before they reported "faster" motion periods, their rate of *microsaccades*—tiny eye movements that occur during visual fixation of an image—increased. Before "slower" or "no" motion periods, the rate of microsaccades decreased. The experiment proved that there is a direct link between the production of microsaccades and the perception of illusory motion in *Enigma*. The illusion starts in the eye, not the brain.

Another of our favorite visual illusions is Mona Lisa's smile. Her expression is often called "enigmatic" or "elusive" but, as our mentor Margaret Livingstone at Harvard University observed, the illusory nature of her smile is explained when you consider exactly how the visual system works. When you look directly at the Mona Lisa's mouth, her smile is not apparent. But when you gaze away from her mouth, her smile appears, beckoning you. Look at her mouth, and the smile disappears again. In fact, her smile can be seen only when you look away from her mouth. This is due to the fact, mentioned earlier, that each eye has two distinct regions for seeing the world. The central area, the fovea, is where you read fine print and pick out details. The peripheral area, surrounding the fovea, is where you see gross details, motion, and shadows. When you look at a face, your eyes spend most of the time focused on the other person's eyes. Thus, when your center of gaze is on Mona Lisa's eyes, your less accurate peripheral vision is on her mouth. And because your peripheral vision is not interested in detail, it readily picks up shadows from Mona Lisa's cheekbones that enhance the curvature of her smile. But when your eyes go directly to her mouth, your central vision does not integrate the shadows from her cheeks with her mouth. The smile is gone.

Mona Lisa (Leonardo da Vinci)

The Best Illusion of the Year contest, mentioned in the introduction, has been a huge success. You would think that after generations of talented, dedicated, sometimes obsessively driven visual artists and scientists tinkering and laboring at their easels, drafting tables, scratch pads, darkrooms, and PC graphics programs, this particular vein of ore would be all mined out. But it isn't.

Consider the Leaning Tower illusion discovered by McGill University scientists Frederick Kingdom, Ali Yoonessi, and Elena Gheorghiu, which took first prize in 2007.

The two images of the Leaning Tower of Pisa are identical, but to you it seems that the tower on the right leans more. This is because your visual system treats the two images as if they were part of a single scene. Normally, two neighboring towers will rise skyward at the same right angle, with the result that their image

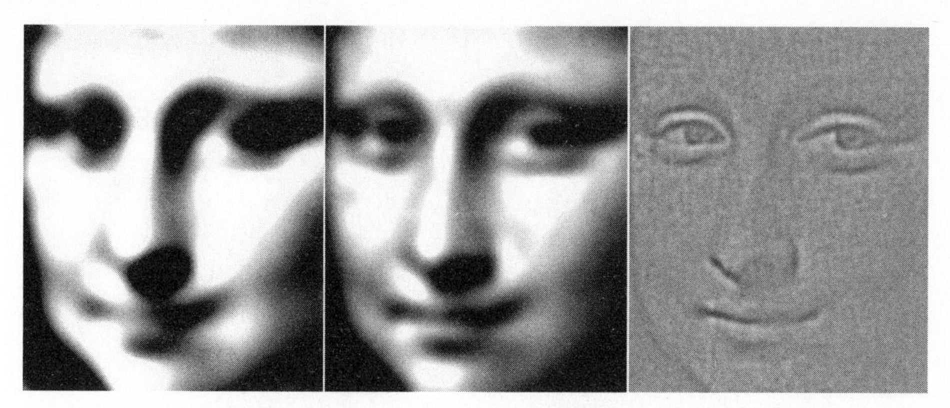

Mona Lisa up close. The three panels are simulations of how your visual system sees Mona Lisa's smile in the far periphery (left), the near periphery (middle), and the center of gaze (right). The smile is more pronounced in the left and middle panels. ("Blurring and deblurring" by Margaret S. Livingstone, Harvard Medical School)

outlines converge as they recede from view. This is one of the ironclad laws of perspective, so invariant that your visual system automatically takes it into account. Since the outlines don't converge in the images above, your visual system is forced to assume that the

The Leaning Tower illusion. (F. A. A. Kingdom, A. Yoonessi, and E. Gheorghiu, McGill University)

two side-by-side towers must be diverging. And this is what you "see."

This illusion is so basic, so simple, it is almost beyond belief that no one ever reported it before 2007. It just goes to show that there is still plenty of low-hanging fruit just waiting to be discovered in the world of illusions. Each new illusion adds depth and definition to perceptual and cognitive theory, bolstering certain hypotheses while weakening others or inspiring new ones. Some suggest new experiments. Each inches us just that much closer to understanding perception and awareness.

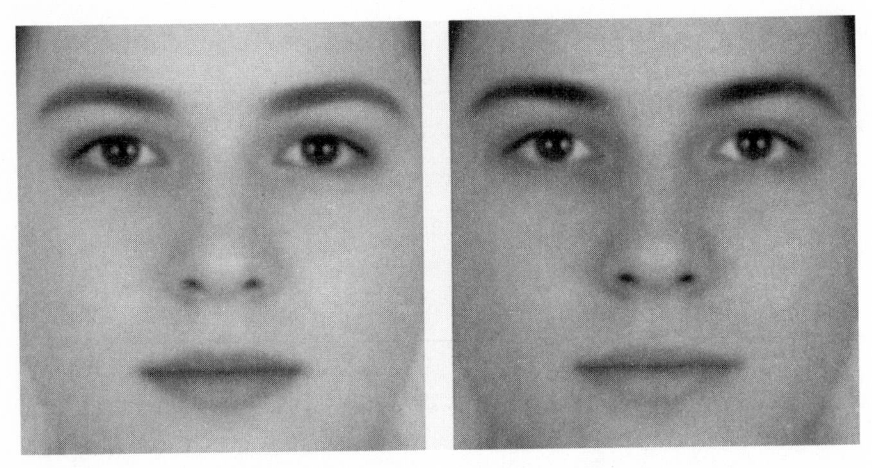

The illusion of sex (Richard Russell)

The only difference between these two faces is their degree of contrast. Yet one appears female and the other male. That's because female faces tend to have more contrast between the eye and mouth (think how makeup exaggerates these features) and the rest of the face than males. Richard Russell, the Harvard University neuroscientist who created the illusion, has found that increasing the contrast of a face (more makeup!) makes it more feminine. Conversely, reducing contrast makes it look more masculine.

Next, the Rotating Snakes illusion, which was presented at the 2005 contest.

The perception of motion need not arise from actual action in the world. Rather, the perception of motion occurs when dedicated

The Rotating Snakes illusion (Akiyoshi Kitaoka)

motion processing neurons in your brain are activated by specific patterns of light intensity changes in your retina.

Some stationary patterns generate the illusory perception of motion. For instance, in this illusion invented by the scientist Akiyoshi Kitaoka, the "snakes" appear to twist. But nothing is really moving other than your eyes. If you hold your gaze steady on one of the black dots in the center of each "snake," the motion will slow down or even stop. Because holding the eyes still stops the illusory motion, eye movements must make the snakes twist. This is supported by the fact that the illusory effect is usually stronger if you move your eyes around the image.

Finally, there is the Standing Wave of Invisibility illusion, which we hope to turn into a totally new magic trick and someday in the future unveil at the Magic Castle. This is the illusion Steve discovered while working on his thesis in graduate school. He wondered what is required for an object to be visible. You might think that visibility should require only that light fall on your retina. But it can be more complicated. Illusions of invisibility show that a stimulus can be projected onto your retina and nevertheless be wholly or partly invisible.

A classic example is *visual masking.* In this illusion, a visual target—for instance, a black bar against a white background—is rendered invisible when two abutting black bars appear a tenth of a second after the target. What's cool is that a target that is seen initially by the brain can be erased by a mask that enters the brain afterward.

Steve's graduate thesis showed how the illusion works in the brain. As it turns out, the target causes two responses in your visual pathway. One, the *onset response*, occurs after the target turns on. A second, the *after discharge*, occurs after the target turns off. Other labs had ignored the after discharge because it occurs after the stimulus turns off. But Steve showed that if you inhibit the after discharge, the stimulus disappears. The same also happens if you inhibit the onset response but not the after discharge. So both the onset response to a stimulus and the after discharge contribute to the neural representation of a stimulus. He realized that if this was true, we should be able to predict a new and very powerful illusion in which a flickering target is perpetually rendered invisible by inhibiting both the onset response and the after discharge of each flicker. It worked!*

We called the new illusion the Standing Wave of Invisibility, and it unites our interest in visual illusions and magic. It is this illusion that we plan to turn into a new stage effect to wow magicians with the power of neuroscience on their own turf. To make this happen we are going to need the help of a magic studio that specializes in electrically engineered lighting effects. For now the trick is on our "to do" list.

*http://sleightsofmind.com/media/standingwave.

4

WELCOME TO THE SHOW BUT PLEASE LEAVE ON YOUR BLINDERS

Cognitive Illusions

Apollo Robbins is sweeping his hands around the body of the fellow he has just chosen from the audience. "What I'm doing now is fanning you," the master pickpocket from Las Vegas informs his mark, "just checking to see what you have in your pockets." Apollo's hands move in a flurry of gentle strokes and pats over the man's clothes. More than two hundred scientists are watching him like hawks, trying to catch a glimpse of fingers trespassing into a pocket. But to all appearances this is a perfectly innocent and respectful frisking. "I have a lot of intel on you now," Apollo continues. "You scientists carry a lot of things."*

Apollo is demonstrating his kleptic arts to a roomful of neuroscientists who have come to Las Vegas for the 2007 Magic of Consciousness symposium. The idea behind this evening is to show these researchers that magicians have much to teach them about the subjects of their life's work: attention, perception, and even the holy grail, consciousness. Magicians and neuroscientists share a passion

*http://sleightsofmind.com/media/magicsymposium/Apollo/.

for understanding the nuts and bolts of the human mind, but we have been developing our respective arts and theories more or less independently of each other for generations. Starting tonight, if all goes as planned, our two communities are going to pay close attention to each other's discoveries.

Apollo has dared everyone in the auditorium to try to catch him pilfering this man's belongings up on stage in plain view. We watch intently just like everyone else, but none of us really stand a chance. This is Apollo Robbins, the infamous "Gentleman Thief" who once pickpocketed ex-president Jimmy Carter's Secret Service detail, relieving them of their watches, wallets, badges, confidential itinerary, and the keys to Carter's limo. He can keep the joke on us for as long as he feels like it, but at least we know one thing he doesn't. As soon as we see who Apollo has plucked randomly from the crowd, we exchange amused glances. This man isn't a scientist at all, as Apollo assumes, but the *New York Times* science reporter George Johnson, who will be explaining to the wider world what transpires here tonight. George is a man of great humor and intelligence, but he is quite shy. His awkwardness makes for great theater.

The fanning continues as Apollo engages in his highly honed rapid-fire patter. "You have so many things in your pockets I'm not sure where to begin. Here, was this yours?" he asks, thrusting something into George's hand. George frowns down at it. "You had a pen in here," Apollo says, opening George's breast pocket, "but that's not what I was looking for. What's in that pocket over there?" George looks over. "There was a napkin or a tissue, maybe? You have so many things it's confusing to me. You know, to be honest I'm not sure that I've pickpocketed a scientist before. I've never had to do indexing as I went through someone's pockets."

Patter, it turns out, is one of the most important tools in the magician's toolkit for attention management. There are only a dozen or two (depending on whom you ask) main categories of magic effects in the magician's repertoire; the apparent wide variety of tricks is all in the presentation and details. Sleight of hand is of course critical to a pickpocket, but so is patter—the smooth and confident stream of commentary that can be used to hold, direct, or divide attention.

Apollo tells George one thing while doing two other things with his hands. This means that in the best-case scenario, George has only a one in three chance of noticing when something of his gets snatched. His real chances are actually far below one in three: in the psychic sparring ring of attention management, Apollo is a tenth-degree black belt. By continually touching George in various places—his shoulder, wrist, breast pocket, outer thigh—he jerks George's attention around the way a magnet draws a compass needle. While George is trying to keep track of it all, Apollo is delicately dipping his other hand into George's pockets, using his fast-driving voice to help keep George's attention riveted on Apollo's cognitive feints and jabs and away from the pockets being picked.

SPOILER ALERT! THE FOLLOWING SECTION DESCRIBES MAGIC SECRETS AND THEIR BRAIN MECHANISMS!

Apollo steals George's pen, notes, digital recorder, some receipts, loose cash, wallet, and, very early on, his watch. One classic way to lift somebody's watch is to first grab his wrist over the watchband and squeeze. This creates a lingering sensory afterimage. You know about visual afterimages from chapter 1—the red dress, the vanishing coin—but afterimages can occur in any sensory system. Apollo is exploiting the same principle, only in this case the afterimage is tactile. The afterimage renders the touch neurons in George's skin and spinal cord less sensitive to the watch's removal and creates a conveniently lasting perception of the watch long after it has disappeared. George simply doesn't notice his watch is missing because his skin tells him it is still there. We notice the watch when we see Apollo folding his arms behind his back, buckling it onto his own wrist as his patter leads George down some new garden path of attention.

END OF SPOILER ALERT

ON ADAPTATION

At one point or another in your life you surely tore your living space apart in search of your glasses—"They can't have just disappeared!?"—only to realize that you were wearing them. When you first put them on an hour ago, the touch receptors in the skin of your face and head gave you a rich sensory impression of their location, their weight, their tightness against your temples. But since then they have become an ineffective stimulus and you feel nothing.

Or try to touch the elastic band of your sock without looking, while you keep your legs and feet still. Chances are you will miss it by at least a couple of inches. This same elastic band was very noticeable against your skin when you first put your socks on this morning. But because nothing has changed since, it has become undetectable to your touch sensors. Or put your hand on a table and hold it completely still. At first you will feel it; after a short time, you no longer notice it.

Adaptation is a critical and ubiquitous process in the nervous system, not just in sensory processing but in all brain systems. It saves energy by reducing the metabolism in neurons that do not receive new information.

A few times during the fleecing, Apollo holds a pilfered object high up behind George's head for the audience to see. This makes everyone laugh but George, who smiles and looks around sheepishly, wondering what the joke is. Then, to more laughter, Apollo returns all of George's belongings one by one. "If you're recording, I think we have evidence," he warns as he hands over the digital recorder. Proffering a folded stack of bills, he says, "I presume this is your gratuity money?" Finally he turns to George and says, "We all pitched in to buy you a watch, very similar to the one you were wearing when you got here." He unstraps George's watch from his own wrist and passes it over. George gasps and then rolls his eyes.

How could George be so inattentive? Why can some joking thief manipulate his attention like a matador leading a bull? It's truly amazing that this can happen to a professionally trained observer like George while he's onstage (and therefore has heightened attention) and has been told what is about to happen to him. It makes you wonder, what is attention? Can you look directly at something and literally not see it?

Magicians are masterminds of human cognition. They control very sophisticated cognitive processes, such as attention, memory, and causal inference, with a bewildering combination of visual, auditory, tactile, and social manipulations. The cognitive illusions they create, unlike the visual illusions discussed so far, are not sensory in nature. Rather, they involve higher-level brain functions. By toying with your cognition—even if they don't know which neural circuits they are tapping—magicians make it impossible for you to follow the physics of what is actually happening. They leave you with the impression that there is only one explanation for what just happened: pure magic.

Possibly the best definition of attention was put forth in 1890 by William James, author of *The Principles of Psychology* and the philosopher king of modern psychology. He wrote: "Everyone knows what attention is. It is the taking possession by the mind, in clear and vivid form, of one out of what seem several simultaneously possible objects or trains of thought. Focalization, concentration, of consciousness are of its essence. It implies withdrawal from some things in order to deal effectively with others."

James elegantly describes the phenomenon of attention, but he says nothing about how it is generated by your brain or how it is modulated in everyday experience. In William James's day, attention could be studied only in terms of introspection—the reflective looking inward on your own thoughts and feelings.

For the next one hundred years, researchers groped in the dark for new and better ways to understand attention. In experiments, subjects wore headphones that piped different words into their left ear and right ear and were asked to listen to just one side, to see if attention could be divided. Some scientists studied radar operators and combat pilots to see how well they could split attention. Others examined the "cocktail party effect," which enables you, in a noisy ballroom filled with loud inebriated people, to hear your name spoken from across the room.

But such studies were observational, meaning the brain was still a black box. Neuroscientists could examine the brain's mechanisms of attention in animals, or in human patients undergoing neurosurgery for diseases such as epilepsy, but there was simply no way to probe the inner cogs and wheels of the brain's attentional circuitry in healthy humans. That changed in the 1990s with the advent of modern brain imaging techniques that allow us to peer into the black box and look for the location of neural correlates of attention. Now we can also begin to figure out how magicians twiddle your attentional circuits with such consummate skill.

Already neuroscientists have learned that attention refers to a number of different cognitive processes. You can pay attention to your TV show voluntarily, which is one process (top-down attention), or your baby's crying can draw your attention away from the TV, which is a different process (bottom-up attention). You can look right at what you are paying attention to (overt attention), or you can look at one thing while secretly paying attention to something else (covert attention). You can draw somebody's gaze to a specific object by looking at it (joint attention), or you can simply not pay attention to anything in particular. Some of the brain mechanisms controlling these processes are beginning to be understood. For example, you have a "spotlight of attention," meaning that you have a limited capacity for attention. This restricts how much information you can take in from a region of visual space at any given time. When you attend to something, it is as if your mind aims a spotlight onto it. You actively ignore virtually everything else that is happening around your spotlight, giving you a kind of "tunnel vision." Magicians exploit this feature of your brain to maximum effect.

It is not yet clear whether there is a single center in your brain that controls attention. Given how many types of attentional effects there are, multiple attention control centers may work in concert. One critical clue is that many of the same brain circuits that control your eye movements are involved with changing the location of your attention in the world. This makes sense, because eye movement circuits are responsible for orienting your eyes to specific areas of visual space, and it seems logical that those same circuits could operate to orient your attentional spotlight, too. Determining what's interesting in the world with attention is undoubtedly critical to deciding where you should look next. Magicians intuitively grasp this, and they control your eyes and your attention as if they were marionettes on a string.

As mentioned, humans have the capacity for *overt* and *covert* attention. When a soccer goalie watches a soccer ball fly toward the goal, she is overtly attending to the ball. But that cagey forward on the opposing team, who's trying to make a shot toward the goal, may intentionally divert the goalie's attention from the ball by looking away from the goal (as if to nonverbally communicate, "Hey, look! I'm going to go over there next!" when in fact the next turn will be in the opposite direction). The move is called a "head fake" in sports, and the idea is to trick the goalie into directing attentional resources to the wrong location. The forward, all along, may have looked toward the fictitious region of interest, but was instead covertly attending to the goal so as to plan her shot.

Too much attention can be a bad thing, too. As social beings, humans and other primates often have to process visual information without looking directly at each other, which could be interpreted as a threat. For example, we all intuitively know not to walk up to a cop, look him or her in the eye, and say, "Hey, what you looking at? You looking at me?" The ability to attend covertly stems from the social circumstance that we do not always want people we are watching to know that we are attending to them.

You also have the ability to engage in *joint attention*. You can gaze at another person, wordlessly pointing to an object with a simple gesture (including a shift in your gaze). By doing so, you may induce that person to look over at the object overtly, or you may induce them to covertly attend to that object. Likewise, when the soccer forward faked out the goalie, she did so by pretending to pay attention to an irrelevant section of the field. She initiated joint attention. Babies as young as nine months display joint attention, as do great apes. Dogs are even better than chimps at some forms of joint attention. A dog will look in the direction you point to. A chimp will not.* Apollo the Gentleman Thief could write the playbook on how to commandeer joint attention.†

A FAILURE OF JOINT ATTENTION

In March 2009, we went to Muhlenberg College, in Allentown, Pennsylvania, to attend the Theory of Art and Magic workshop. Each day of the workshop was filled with theoretical lectures, hands-on seminars, and performances. We witnessed a virtuoso performance by Roberto Giobbi of Switzerland, who also gave a full-day workshop on card tricks that complemented his highly regarded five-volume set *Card College.* (So when we say Roberto wrote the book on card tricks, he really wrote five.)

We were sitting in what clearly used to be an upscale private home, now used by Muhlenberg College to host small conferences and meetings with donors. Roberto worked miracle after miracle, and then he performed his version of the famous Bitter Lemon trick. In this trick, a magician asks a spectator to pick a card and sign it, only to find that the card has been transported to the inside of an uncut piece of fruit. The

*To investigate joint attention, researchers hide an object or food inside an opaque container and point at it with their finger. Babies fourteen months and older are able to locate the hidden item easily, but chimps find this task extremely challenging. Surprisingly, domestic dogs excel at solving the same problem. References: B. Hare et al. (1998), "Communication of Food Location Between Human and Dog (*Canis familiaris*)," *Evolution of Communication* 2: 137–59; A. Miklosi et al. (1998), "Use of Experimenter-Given Cues in Dogs," *Animal Cognition* 1: 113–21.

†Joint attention permeates every social interaction, in sophisticated and often subtle ways. Neuroscientist Sonya Babar and her colleagues found that when we look at somebody's face, we shift our gaze between the two eyes of our partner, seeking the best eye contact. The eye we settle on at any given time tends to be the mirror image of the eye chosen by our interlocutor. For instance, if we sense that our partner's eyes are focused on our right eye, we reflexively respond by shifting our gaze to her left eye. This joint shifting of gaze is perceived as proper eye contact. However, if a partner looks at our right eye as we look at her right eye, both of us will usually sense a break in eye contact or lack of attention. Reference: S. Babar et al. (2010), "Eye Dominance and the Mechanisms of Eye Contact," *Journal of AAPOS* 14: 52–57.

fruit is given to the spectator along with a knife, and when she cuts it open, she finds a rolled-up card. You guessed it—it's hers.

Roberto, a traditionalist, actually uses a lemon. But his trick adds a twist. In his version, Roberto lays a handkerchief over his empty hand and the lemon appears under the handkerchief as if from nowhere. It's a beautiful sleight that fooled everybody in the room. Except Susana.

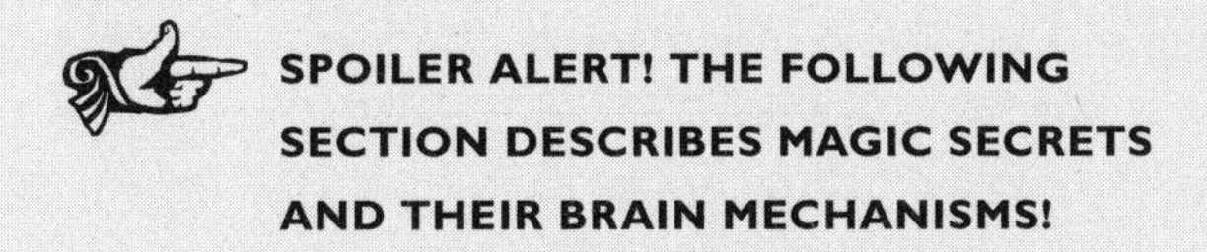

SPOILER ALERT! THE FOLLOWING SECTION DESCRIBES MAGIC SECRETS AND THEIR BRAIN MECHANISMS!

Susana, you see, was pregnant with our second son, Brais, and suffered from all-day morning sickness during Giobbi's workshop. She wasn't paying attention. Whereas Giobbi had the rest of the crowd concentrating, Susana was busy trying not to puke. Then a flash of yellow finally caught her attention. She gazed at the magician to see, plain as day, that he was stuffing a lemon up underneath the handkerchief into the palm of his other hand. Later, she mentioned to Steve that she thought that trick was uncharacteristically sloppy. She didn't understand why the professional magicians attending the class were so amazed. Steve had no idea what she was talking about. He thought the lemon sleight had been seamless. That's when Susana realized that she had been able to detect the method behind the trick only because of her queasiness-induced attention deficit. Roberto controls people's perception by focusing their attention on his face just as he unceremoniously shoves the lemon up under the handkerchief. This is joint attentional control at its finest. But Susana's attention was fully focused on her barf control mechanism, and thus was unmanageable even by a master magician.

END OF SPOILER ALERT 

Attention is also linked to your short-term memory and your ability to tune out what is happening around you. Sometimes a stimulus is so demanding, so salient, that you cannot help but pay attention—an ambulance siren, an infant's cry, a dove fluttering out of a top hat.

This information flows in a bottom-up fashion—from your primary senses to higher levels of analysis in your brain. It is called *sensory capture.*

Other times you can shift your attention around, as you choose, in a top-down fashion. Signals flow from your prefrontal cortex (the CEO of your attentional networks) to other regions that help process information. This is the spotlight of attention that is under your control. You don't hear the siren or the baby or see the dove because you are attending to something else, such as the last page of that gripping mystery novel you are reading. Research shows that the greater your capacity for short-term or working memory, the better you are at resisting sensory capture.

Neuroscientists have begun to dissect the nature of attention and identify its neural correlates. The initial brain areas that process a visual scene use circuits that lay out visual space like a map. These first few stages of visual processing (the retina, the visual thalamus, and the primary visual cortex discussed in chapter 1) are organized so that the neurons that process one part of the visual field are positioned directly next to neurons that process the adjacent parts of the visual field. As your eyes move around, your retinas and the visual input move around, too. But no matter where you look, some neurons are assigned to your central vision, and the other neurons are assigned to specific peripheral positions of input from your retinas. The retinal positions of these visual neurons never change.

When you decide consciously to pay attention to a specific location of this "retinotopic" space, neurons from higher levels of your visual system increase the activation of the low-level circuits and enhance their sensitivity to sensory input. At the same time, neurons in the surrounding regions of visual space are actively inhibited. We recently worked with a group of colleagues led by neuroscientist Jose-Manuel Alonso at the State University of New York and showed that the neurons in the primary visual cortex not only enhanced attention in the center of the spotlight and suppressed attention in the surrounding regions, but their degree of activation was modulated by the amount of effort used to accomplish a given task. In other words, the harder the task, the more the central region of attention was activated and the more the surrounding region was suppressed.

In a magic show, you face an incredibly difficult task: to peel away all the layers of misdirection and figure out the secret method underlying each magic effect. But the harder you try, the harder it gets: the more your attention is enhanced on the center of the attentional focus, the more your attention is suppressed in all other locations. Of course, the center of the attentional focus is right where the magician wants it—where nothing of particular interest is going on. The locations surrounding your spotlight of attention—where the real action is happening—are now conveniently suppressed by your brain. The armies of neurons that suppress perception in those regions are the magician's confederates.

Apollo works his marks as if he knew about these neuronal circuits all along. He'll pull a quarter from your breast pocket and ask, "Is this yours?" You know full well that it's not yours (who keeps quarters in their breast pocket?). But you can't help it, you inspect George Washington's face as if you might find your initials engraved on his forehead. "What year is the coin?" Apollo asks. And you dutifully try to make it out, but the letters are too small and blurry, so you reach for your reading glasses . . . in your breast pocket. They are missing. "Try these glasses," Apollo kindly offers as he hands you the glasses off his face. Your own glasses, as it turns out. While you were busy attending to the quarter, which you knew didn't actually come from your pocket, Apollo's hands absconded with those glasses literally right under your nose while you suppressed all visual motion surrounding the quarter.

If neuroscientists had known—as Apollo seems to know—that attention works in this way, it would have saved a whole lot of research time. So now we study magicians.

ON MISDIRECTION

You don't have to be a magician to be skilled at attentional misdirection. When a conversation edges into uncomfortable territory, your natural instinct is to change the subject. Often the other person plays along, as if you weren't just talking about your testicular cancer, and pretends that yes, we really are talking about last night's Red Sox score. Our brains are designed to be flexible with regard to what we are paying attention to, at both the sensory and the cognitive levels. Without this flexibility we would be unable to drive home thinking about what's for dinner and then instantaneously swerve the car to avoid the child chasing her ball into the street.

After fleecing George, Apollo turns to the audience and says, "Now would you like to see the behind-the-scenes of how I did all that?" Magicians are famously loath to give away their secrets, but Apollo is here in Las Vegas tonight to instruct, not just to entertain. He calls the ever amiable George back for more pilfering, but this time he explains what he is doing. He slows his techniques way down, occasionally pausing and rewinding.

Most people call what magicians do "misdirection," explains Apollo, but that is like saying doctors make people well with their curing skills. The term is so broad that it is next to meaningless. He prefers to discuss specific principles and techniques such as "frames" and "attention management." It's not true, he says, that the hand is quicker than the eye. Most manipulations are carried out at a normal pace. Success relies on the magician's skill in diverting your attention away from the method and toward the magical effect.

SPOILER ALERT! THE FOLLOWING SECTION DESCRIBES MAGIC SECRETS AND THEIR BRAIN MECHANISMS!

Frames are windows of space that the magician creates to localize your attention. A frame can be the size of a whole room or a tabletop or no bigger than a business card. "You have no choice but to watch in the frame," says Apollo. "I use movement, context, and timing to create each frame and control the situation." Apollo demonstrates by moving very close to George. He grabs George's hand and pretends to press a coin into it, though all he is really placing there is another sensory afterimage with his thumb. "Squeeze hard," says Apollo. George gazes intently at his hand, now caught within a frame. He squeezes. "Do you have the coin?" teases Apollo. George nods. He thinks so. "Open your hand," says Apollo. The palm is empty. "Look on your shoulder," says Apollo. George glances to his shoulder, where a coin is resting.

Apollo explains that if a subject's attention is localized to a frame, then maneuvers outside the frame will rarely be detected (such as placing a coin on a shoulder). Magicians, he says, thoroughly manage attention at all times. People tend to think of misdirection as the art of making someone look to the left while some fast move is pulled on the right, but Apollo says it is more about force-focusing your spotlight of attention to a particular place and at a particular time.

Magicians exploit several psychological and neural principles to focus your spotlight of attention. Recall that when you see an object that is new, bright, flashy, or moving—think of that white dove fluttering out of a top hat—your attention is driven by increased activity in your ascending sensory system, which simply means that salient information from your senses flows up into your brain. It arrives from the bottom and travels up. You are strongly drawn to the object. Neuroscientists call it *sensory capture.* Psychologists call it *exogenous attentional capture.* Magicians call it *passive misdirection.*

In passive misdirection, you are attending to the fluttering bird while the magician gains a few unattended moments to carry out a sneaky maneuver. It is passive because the magician lets you do all the work. He just sets up the condition. In Penn & Teller's version of the cups and balls, Penn uses his juggling skills to draw your attention while Teller does a secret move. Penn actually tells you what he's doing. "This is not juggling," he says as the three little aluminum foil balls cycle in front of his face, "this is misdirection." You of course helplessly watch the juggling show intently right up until the point Penn informs you that you've been duped.

If more than one movement is visible—the flying dove arcs overhead while the magician reaches his hand into a box to set up the next trick—you will naturally follow the larger, more salient movement. You track the bird, not the hand. Hence the magician's axiom, "A big move covers a small move." In fact, a large or fast-moving stimulus, such as the fluttering dove, can literally decrease the perceived salience of a small or more slowly moving stimulus, such as the magician's hand in the box, so that your attention is drawn to the bird, not the hand. You already know the reason: when you pay attention to a particular location in space, the neurons responsible for processing information in the surrounding regions are inhibited.

When two identically salient actions start simultaneously, the one you notice first captures your attention. It not only becomes more salient, but the other action is suppressed, becoming less salient. Furthermore, things that are novel (the unexpected dove) produce stronger responses in parts of your brain that are critical to the allocation of attention (namely, the inferotemporal cortex, the hippocampus, the superior colliculus, the prefrontal cortex, and the lateral intraparietal area; these areas receive the bottom-up sensory signals and then activate circuits that enhance the attended object while suppressing other objects in your visual field). The salience of an object also increases when a magician actively directs your attention to it. For example, Apollo may ask you to leaf through the pages of a book while he places your stolen wallet in his pocket. You become absorbed in the task of turning pages. This is *active misdirection.* (Psychologists call it *endogenous attentional capture.*) Your top-down attentional control is focused on the book and you ignore the hand. The magician's actions enhance the firing of neurons involved in your attention to turning the pages of the book, whereas neurons that might attend to the magician's hands are suppressed.

Apollo messes with your head in other ways as well. His patter aims to generate an internal dialogue in your mind—a conversation with yourself about what is taking place. This, he says, results in a great deal of confusion. It slows your reaction time and leads you to second-guess yourself.

Many magicians use comedy and laughter to reduce your focused attention at critical points in their acts. Remember the Great Tomsoni and his corny jokes? He takes advantage of your diminished attention in those offbeat moments when you relax after a joke. Or Magic Tony and his leopard shoes? Tony's magical patter aims toward plays on words and homespun rhetoric. He's created a character who fully embodies one of the primary stereotypes of a magician: the corny joking uncle.

Tony says that his goal is for his patter to be "so lame that it's cool." We couldn't help but wonder why he chose a persona that is, well, overwhelmingly lame. Tony says that the accidental cornballs create an atmosphere in which you may laugh at their jokes, but it's because you feel you have to be polite, not because it's funny. Without the fake laughter, the show would be embarrassing for everyone, so you laugh. But Tony realized that an unrepentant, over-the-top, intentionally corny punster can make you into a willing executioner of his humor. And that can be very useful to him as a vehicle for misdirection. An honest groaning response to a pun is more attention-grabbing than a fake laugh, says Tony. It's hard to stay focused on the method of a trick when you're busy cringing or rolling your eyes.

In many magic tricks the secret action occurs when you think that the trick has not yet begun or when you think that the trick is over. Magicians call this *time misdirection.* They can also introduce delays between the method behind a trick and its effect, preventing you from causally linking the two. Arturo de Ascanio, the great magic theorist and father of Spanish card magic, refers to this specific type of time misdirection as the "parenthesis of forgetfulness." Essentially, it means that the magician must separate the method from the magical effect. This separation messes up the spectators' reconstruction process.

Imagine that a magician fakes a coin transfer from his left to his right hand, and then opens his right hand to reveal that it is empty. Because there is no separation between the sleight (the fake transfer) and the magical effect (the vanished coin), you may easily conclude that the coin was never actually transferred but remained concealed in the magician's left hand. A more accomplished magician will introduce a separation—a parenthesis of forgetfulness—between the method and the effect. For example, after the fake coin transfer, and before revealing his empty right hand, he may reach into his pocket for the overt purpose of retrieving a magic wand, but in fact he is also dropping the palmed coin inside his pocket. Then, touching the magic wand in his left hand to his right hand, he shows that the coin has disappeared. When you rewind the scene in your mind, you will have a harder time figuring out where the vanished coin might be hidden.

One of Magic Tony's tricks involves misdirection based on what psychologists call a *habituation-dishabituation paradigm*. This means he specifically tries to make you complacent (that is, bored, lazy, or otherwise not carefully attending to what he is doing) by apparently repeating the same action over and over, and to lull you into a false sense of security. That's habituation. And then *bam!* he changes the method, leading you to the resultant spectacular effect.

Nobel laureate Eric Kandel and our friend Tom Carew showed that one of the neural correlates of habituation-dishabituation is a change in the strength of connections between neurons in your brain. When habituation occurs, neurons send less signaling chemicals (neurotransmitters) to the neurons they are connected to, thus decreasing the response downstream. When the same connection becomes dishabituated, the signaling neuron sends lots of neurotransmitters once again, thus restoring the bigger reaction in the downstream neuron. Tony elegantly switches the audience's neurons from habituation to dishabituation modes. His initial repetitions lull the spectators' brains into mind-numbing habituation, only to be brusquely awakened (dishabituated) by the shocking magic effect he finally achieves.

END OF SPOILER ALERT

Another important concept, Apollo tells the scientists gathered in Las Vegas, is that tricks are embedded in natural actions. He demonstrates by making a pen disappear. He dangles it in front of the audience with one hand. When he flicks his other hand past his ear, as if to scratch, no one notices. The movement is natural, unremarkable, quick. Suddenly everyone sees the pen has vanished. Apollo turns his head around to reveal the pen tucked behind his ear.

Teller, the shorter half of the duo Penn & Teller, sheds his mute persona to describe the same concept. A former high school Latin teacher, Teller is far from mute offstage. He has a love for words, and his explanations are not only scholarly but unexpectedly eloquent. "Action is motion with a purpose," he says. In normal social interactions, we constantly search for the purpose motivating other people's actions. An action with no obvious purpose is anomalous. It draws attention. However, when the purpose seems crystal clear, we look no further. Teller explains that he will draw suspicion if he raises his hand for no apparent reason, but not if he performs a seemingly natural or spontaneous action like adjusting his glasses, scratching his head, draping his coat over the backrest of a chair, or reaching into his pocket for a magic wand. Teller calls this "informing the motion." He says, "Skilled magicians inform every necessary maneuver with a convincing intention."

Neuroscientists now have a good idea why such decoy actions are so good at fooling us. It comes from a remarkable type of brain cell called a *mirror neuron.* You are familiar with the idea of the "mind's eye": pretty much at will, you can conjure a quasivisual experience of just about anything that can be seen or depicted in images. You also have your "mind's ear," with which you can replay songs and noises and voices you are familiar with. Similarly, there is your "mind's body." This is your brain's virtual representation of your physical self. When you plan out how you are going to cook tonight's dinner, when you daydream that you are an action hero, whenever you relive a painful memory of gym-class humiliation, you are running a virtual simulation of those actions in your mind's body. It is an invaluable psychic tool for planning and executing actions, learning motor skills and remembering them. Mirror neurons form an important part of your mind's body because they help you understand the actions and intentions of other people. They do this by automatically mimicking others' actions and assuming their intentions using your own mind's body. So when you see Teller reach for a glass of water, you instantly do the same thing in your mind's body. You also ascribe a simple, natural motivation to him, namely, that he is thirsty and will raise the glass to his lips and take a drink. In your mind's body, you do this, too. Literally: many of the same neurons that are active when you take a drink are active when you think someone you can see is about to take a drink. Your brain makes a prediction and runs a simulation, automatically and usually subconsciously.

Mirror neurons are an important element of human social intelligence. They are part of how we are able to understand each other, to imitate, to learn and teach, to empathize. But they can also mislead us. A good magician can disguise one action as another or convincingly fake an action he isn't really performing, prompting your mirror neurons to feed you false inferences about what he is actually doing or not doing. You see Teller raise the glass to his lips and seem to drink, and your automatic prediction seems to be fulfilled. But did he really take a drink? Maybe he transferred something from hand to mouth, or from mouth to hand.

ON AUTISM

Joint attention is the mechanism by which you can share another person's experience by following the direction of his or her gaze and pointing gestures. A common and medically established symptom of many autism patients is that they have a deficit in joint attention which can be measured by tracking their eye movements. For instance, autistic patients tend to not look at other people's faces, even the faces of actors in movies or subjects in photographs.*

Magicians rely on joint attention as a form of social misdirection, to divert your attention from the method behind the trick and toward the intended perceptual effect. If the magician wants your eyes focused on his face, he will look directly at you. If the magician instead wishes you to shift your gaze to a particular object, he himself will turn his body, head, and eyes toward that object, and your head and eyes will quickly follow. This is the magician manipulating your joint attention. In a double act such as Penn & Teller's show, the opportunities to capitalize on joint attention increase twofold. When Penn Jillette performs a routine, Teller's body, head, and gaze are intently oriented to the location of attention the duo wishes to impose on the audience (Penn's hands, face, a specific object onstage) and vice versa. We were careful to apply this same principle when practicing our joint act for the Magic Castle. Joint attention is critical for language acquisition and cognitive and social development. But it also makes you susceptible to magic tricks that exploit your natural impulse to pay attention to the same places and objects attended to by the people around you.

Our hypothesis is that autism patients who suffer from problems of

*See "Eyes: A New Window on Mental Disorders," *Scientific American*, February 2009.

joint attention should respond abnormally to magic tricks that rely on joint attention. They will not be duped by social misdirection, so they will be more likely to "catch" the magician's secret action than normal observers. Failure to be fooled by magic tricks that rely on social misdirection would thus indicate that joint attention is impaired, which could help the diagnosis of autism-spectrum disorders. It would also help evaluate the success of therapies directed to improving joint attention: as the patients' joint attention improves, they should become more and more susceptible to social misdirection and thus more likely to "fall" for magic tricks that rely on joint attention cues. We have written a grant proposal to fund a study to determine whether our hypothesis is correct.

Unlike people with autism, most of us turn our gaze and attention to the faces of people in photographs. However, our intense focus on faces is at the expense of other potentially interesting information. Have you noticed anything strange about this picture? Look more carefully, and you may see that the girl has an extra finger on her right hand. Observers with autism may be quicker to notice details such as these because their attention is not fixed on the faces. (Photocomposition by Smitha Alampur, Thomas Polen/iStockphoto)

Social misdirection onstage, as used by the magician, is only a more refined form of the social misdirection used by our primate cousins to procure themselves better access to food and other resources. Ethological studies have shown that a macaque will avoid looking at a hidden food cache so as to keep potential competitors away. Consciousness researchers say that such macaques have a *theory of mind.* That is, they know to interpret the gaze, head, and body orientation of their peers as indicators of their location of attention and interest. They also know how to adjust or redirect their own body and gaze to fake interest in an undesired object so as to draw competition away from the object of desire. In this sense, both macaques and humans are proficient mind readers. But magicians are best. And Apollo, as you'll see in the next chapter, has even more tricks up his proverbial sleeve.

5

THE GORILLA IN YOUR MIDST

More Cognitive Illusions

Apollo Robbins is having a blast fleecing George at the Magic of Consciousness symposium. He turns to face him for another demonstration of his wiles. "When I approach somebody," he says, "I find that if I go straight in, I enter their personal space. It's like a bubble surrounding their body. The distance is different in different cultures and in different people, but everyone senses the space and tries to protect it." Apollo then turns his body to stand shoulder to shoulder with George. "But if I move to the side, like this, the gap is much smaller. You don't feel invaded." One more thing. "As I move into your personal space, I need to break eye contact with you, so that you don't keep your gaze on me." Apollo looks down. George looks down. Apollo pops up next to George's shoulder. He is now safely inside George's bubble. He can get away with magical murder.

Apollo's observation is fascinating. What he calls *personal space*, neuroscientists know as *peripersonal space.* (Scientists can never resist a good game of Pin the Greco-Latin Root on the Simple Word.) People have always had a strong intuitive sense of this space, and neuroscience has recently begun to decode its neural foundation in the brain.

It turns out to be more than a mere metaphor but less than a real, tangible aura. It is a construct your brain actively creates as part of your mind's body. As far as your brain is concerned, the space immediately around you is literally a part of your body. This is why you can tickle a child by wriggling your fingers in the air over her ribs, and why you are physically as well as emotionally sensitive when someone "punctures" your bubble uninvited.

Finally, Apollo reveals a principle of the pickpocket's art that particularly thrills us as neuroscientists. "In years of doing shows," he says, "I noticed that the eye is more attracted to arches than to straight lines." He starts patting George's pockets again. George looks on with interest. "If I want to take something out of his pocket, I can keep his eyes occupied on my free hand if I move it in an arc. But if I move it in a straight line, his attention will snap back to my other hand" like a rubber band, he explains.

We had first heard Apollo describe this principle when we came out to Las Vegas a few months prior to the Magic of Consciousness symposium, in one of the meetings where we got together with magicians to share knowledge and ideas and to brainstorm about the upcoming conference. (We don't mind telling you that after every meeting with Apollo we check the credit cards in our wallets to see if they've been swapped for fakes. He's really that good.) Teller had called for this particular meeting in his office so that we could present to the magicians our scientific research on illusions and visual perception. The initial purpose of our collaboration with magicians was to enable us to use magic in the lab, but it would obviously help for the magicians to know what cognitive research looks like. After showing them some of our work on visual illusions, Susana presented what we know about the neuroscience of eye movements. There are two main kinds, and they serve different purposes and are probably controlled by different subsystems of the oculomotor system.

In the first kind of eye movement, called *saccade*, your eyes jerk almost instantaneously from one point to another. The fleeting moments between saccades, when your eyes are for the most part motionless, are called *fixations*. Saccades are critical to vision because your eyes can make out fine detail only in a keyhole-sized circle at the very center of your gaze covering one-tenth of one percent of your retina; the vast majority of the surrounding visual field is of shockingly poor quality.

You can prove this to yourself with an ordinary deck of cards. Separate out the face cards and shuffle them. Fix your gaze on something directly across the room and don't let your eyes move at all. Draw a random face card and hold it out at arm's length at the very edge of your peripheral vision, then slowly pivot your arm forward, bringing the card toward the center of your unflinching forward gaze. Assuming you can resist the urge to let your eyes dart off to steal a glimpse, you will find that the card has to come quite close to your center of vision before you can identify it.

The reason it doesn't *feel* like your vision is ninety-nine point nine percent garbage is because of saccades. Your eyes are constantly darting around the world like a hummingbird on meth. Your brain edits out the motion blurs and integrates the small bits of information received from each fixation in order to present your visual awareness with a detail-rich, stable-seeming portrait of the visual scene before you.

Saccades are also related to adaptation. Recall that the neurons in your visual system are designed to detect change. But when conditions remain static, your neurons adapt by slowing their firing rate. They cease giving you reliable information, and your perceptions are limited. It is as if your neurons actively ignore a constant stimulus to save energy so as to better signal that a stimulus is changing. The visual scene threatens to fade away.

To overcome adaptation, you make microscopic eye movements during each fixation between large eye movements. Such fixational eye movements are essential for vision. Indeed, without these minuscule ocular meanderings, you would be blind when you fix your gaze. Our studies indicate that when your gaze stops on an object and does not move, activity in your visual neurons is suppressed. The object disappears!

In the second kind of eye movement, called *smooth pursuit*, your eyes move in a continuous, uninterrupted path without any pauses or jerks along the way. Smooth pursuit takes place only when you track a moving object. It cannot be faked. This is one of the reasons that

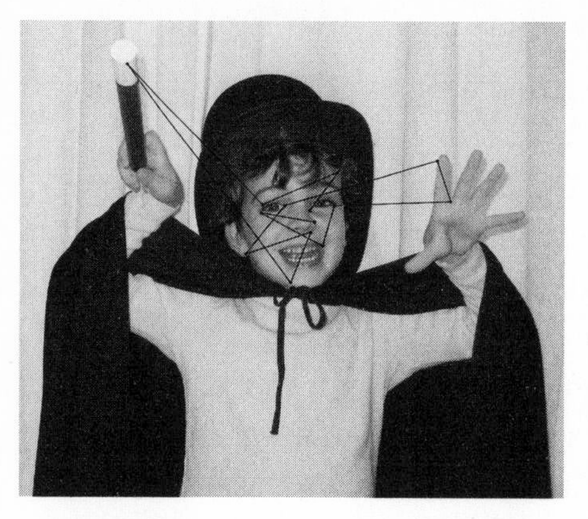

Saccadic eye movement vs. smooth pursuit: the left figure shows the zigzagging path an observer's eyes might trace while looking at a magician. The right figure shows the eyes' smooth, unbroken pursuit path as they follow the tip of his wand raising in a gentle arc. (Photograph by Matt Blakeslee)

some scenes in movies fail: when an actor pretends to track an object that doesn't actually exist, but is added in postproduction, the eye movements inevitably look wrong on-screen. Pursuit eye movements allow you to track moving objects, while saccades systematically search and gather information from a visual scene.

You can observe the difference between these two types of eye movement by holding up your thumbs in front of you about a foot apart. Now, holding your hands still, ask a friend to slowly move her eyes as smoothly as possible from one thumb to the other. Notice that her eyes make little jumps along their journey. Those little jumps are saccades. No matter how hard she tries, she cannot make her eyeballs swivel smoothly between the thumbs. Now try it again, but this time ask her to watch your left thumb as you move it slowly over to touch the right one and then back out again. Notice this time how her eyes track perfectly smoothly.

All the magicians were fascinated by these facts, but for Apollo they triggered a eureka moment. He said that as a pickpocket he differentiates between straight-line and curved hand movements when managing his marks' attention. He now realized the reason might be the difference between saccades and pursuit eye movements.

When you see a hand quickly moving in a straight line, your

eyes—and your attention—automatically jump to the end point. So a pickpocket will make a fast, linear gesture if he wants to minimize your ability to pay attention to the path itself. But a hand that moves in an arc triggers a different tracking mechanism. You cannot predict where the hand is headed, so you fixate on and follow the hand itself, and you fail to notice when Apollo's other hand slips into your pocket.

Pickpockets have a whole toolkit of misdirection techniques. We were already familiar with some of them. Such thieves often ply their trade in dense public spaces and rely heavily on socially based misdirection—eye contact, body contact, and slipping, ninja-like, inside the personal space of the mark. But Apollo's observation was new to us, and it immediately spawned new ideas for experiments.

It is well established that visual perception is suppressed during saccades, which could explain the way pickpockets make use of fast linear movements. But what about attention? Is it also suppressed during eye movements? Scientists do not yet have an answer, but Apollo's suggestion was so intriguing that we wanted to take it to the lab. This conversation marked a sea change in our relationship with the magicians. Our original intention had been simply to poach their best techniques so that we could design better experiments, but now we realized that magicians might actually know things about mind and behavior that neuroscientists do not.

You already know about your capacity for "overt" and "covert" attention. *Overt attention* is when you purposefully direct your eyes to an object while paying attention to it. *Covert attention* is the act of looking at one thing while paying attention to another. Magicians, diabolical as ever, have exploited these properties of your brain in designing some of their favorite tricks. To describe these methods, we coined the terms *overt misdirection* and *covert misdirection.*

In overt misdirection, the magician moves your gaze away from the method behind the trick. He draws your eyes to something of false interest while he carries out a secret action at another location. This is what most people think about when they hear the word "misdirection." An explosion lights up the stage, and a miniature mushroom cloud billows its way up to the rafters. Whoops! Where did that rabbit come from on the other side of the stage? When you were looking at the explosion, the magician used any one of a dozen methods to make the rabbit appear while you were distracted. That's overt misdirection, and it's the same thing Steve did when he swiped Halloween candy as a kid. "Hey, Jimbo! Is that the Goodrich blimp?" Candy gone. And by the time the theft is discovered, it's half eaten. Yes, Jimbo is Steve's younger brother, and this is a fond memory of nutty, chocolatey, stolen goodness.

Covert misdirection is more subtle. The magician draws your attentional spotlight—and focus of suspicion—away from the method without redirecting your gaze. You may look directly at the method behind the trick, but you are entirely unaware of it because your attention is focused elsewhere. You look, but you do not see.

Cognitive neuroscientists know quite a lot about covert misdirection—it's a critical element in inattentional blindness. With inattentional blindness, you fail to notice an object that is fully visible because your attention has been directed elsewhere. It pertains to how your brain sees and processes information. We also study a closely related phenomenon called *change blindness*. With change blindness, you do not notice a change in a scene. It pertains to how your mind fails to remember what it has just seen.

CAN YOU KEEP US FROM READING YOUR MIND?

Can you explain the astounding results of the following mind-reading experiment by Clifford Pickover, a prolific author of popular books about science and mathematics? The editors of *Scientific American* prepared a simulated Pickover test that you can take here, or you can try the even more puzzling online version at http://sprott.physics.wisc.edu/pickover/esp.html.* By using ESP, we think we can predict the outcome of your

*Pickover's Test is based on a trick invented by Henry Hardin, around 1905. Hardin originally marketed it as the Prince's card trick, but over time the title slipped into the Princess card trick, which is now its proper name.

choice with 98 percent accuracy. To begin, pick one of the six cards below and remember it. Say its name aloud several times so you won't forget it. Once you're sure you'll remember it, circle one of the eyes in the row below. Then turn to page 82 to see if we are right.

While many magicians strive to exploit inattentional or change blindness in their acts, the grand master of these deceptions is the Spanish magician Juan Tamariz. In the hierarchy of illusionists, he is Yoda. Dai Vernon, the legendary magician who fooled Houdini (chapter 2), used to say that in his eighty-plus years of career as a magician, nobody had been able to deceive him like Tamariz. But you wouldn't know it from looking at him. Sure, we've discussed some weird-looking magicians. But when you conjure up the image of a world-famous magician in your mind's eye, you nevertheless probably think classy: well dressed, well coiffed, well mannered. You think Copperfield, Henning, even Penn & Teller in their matching suits.

But an unkempt Spaniard with long stringy hair and crooked teeth who wears huge eyeglasses, goofy vests, and a purple top hat? This guy has the propensity, at the climax of a trick, to jump into a Gollum-like, bent-over posture and point at you while he screams *"Chaan ta ta chaaaaaan!"* No one would imagine that this comical Cat-in-the-Hat character would be a top mage—which is one of the primary reasons he is so effective at duping you.

Tamariz is a founder of what is known as the Madrid School of

Magic (Escuela Mágica de Madrid). It's a magic think tank of like-minded conjurers from around the world who are interested in improving the art of magic through the application of human psychology. Members consider every aspect of the art, from the minor issue of which way to reveal a card (it's better to flip it head over heels rather than heels over head*) to the important question of exactly when and when not to introduce humor during a trick.† Their goal is to understand magic methods and the human mind to such a high degree that magic tricks make you feel as though a miracle has just happened.

Tamariz uses inattentional blindness to create many small miracles. He figures that you probably don't know you are blind to things outside your spotlight of attention. So when he performs a trick obviously right before your eyes—and you miss it—you will be incredibly surprised; the only explanation is magic. In one such method, called Crossing the Gaze, Tamariz makes a coin disappear from one hand while keeping both hands openly visible.

WE READ YOUR MIND . . .

We have removed your card! Did we guess the card you picked on page 81? If so, does Pickover's ESP system explain our correct answer, or is there a simpler explanation? Read no further until you want to know the

*Dai Vernon was the originator of the studies on how best to flip a card to reveal it at the end of a trick.

†Juan Tamariz has written and lectured extensively on how to combine magic and humor and on the difficulty of achieving a balance between both.

answer.* Give up? Look once more at the six cards on page 81, then compare them with the five cards pictured on page 82. Notice any differences? If the act of circling an eye distracted you and you fell for the trick (most people do), you are a victim of what psychologists call change blindness. A change—even a big, obvious change—can be all but invisible until you take another look.

*Clifford Pickover's Web site offers some hilarious explanations by people who tried the test at their computers. This is our favorite one. From: Petri Kotro (Finland) University of Lapland. Dear Cliff, your program removed several times the card I named, even though I spoke Finnish when naming the card. There are not many people in the Anglo-Saxon world who can read a Finno-Ugric mind that easy (or know any Finnish). Or maybe you're Celtic?

SPOILER ALERT! THE FOLLOWING SECTION DESCRIBES MAGIC SECRETS AND THEIR BRAIN MECHANISMS!

Here's what the trick looks like. Tamariz stands with his right side facing you. His left hand is outstretched, palm up and empty. His right hand points to his open palm. Tamariz looks at you, beckoning

Tamariz's Crossing the Gaze technique (the gaze motion should cross the hand motion, so that the two trajectories are equivalent but their directions opposite) was inspired by the Italo-Argentinian magician Tony Slydini, one of Tamariz's masters. (Courtesy of Juan Tamariz)

your gaze directly to his eyes. He has your full attention. Then he looks down at his empty palm. You follow his gaze and look at the palm. And here is the essence of the trick. During the fraction of a second while you move your eyes, Tamariz lifts his right hand toward you in a natural gesture that says "Hold on, don't be impatient." And there, in the middle of his right palm, is a bright shiny coin. It's in full view. But you don't see it because Tamariz has powerfully driven your attention to the empty palm. You concentrate so hard that you miss an object that is reflecting photons directly onto your retina.

END OF SPOILER ALERT

So what is the point of this maneuver? You never knew the coin was there, so why is he taking the trouble to misdirect your attention? A good magician can take advantage of this situation in countless ways. For instance, Tamariz can now do something else with his right hand to produce the coin. But you "know" both of his hands were empty because you "saw" them that way. It is this kind of strong, albeit misleading, evidence that will make the subsequent appearance of the coin feel like a miracle.

Neuroscientists are equally thrilled with the possibilities raised by inattentional blindness. Several years ago, two of our colleagues, Daniel Simons and Christopher Chabris, designed a brilliant experiment that never fails to shock and delight people encountering it for the first time. The instructions are simple. You are asked to look at a short video of people passing around a basketball. One team wears white T-shirts, the other wears black T-shirts. Your job is to count the number of passes made by one team, or to keep count of bounce passes versus aerial passes. After three or four minutes, the video ends and you are asked if you saw anything unusual.*

No? Look again. This time the scientist pauses the video at the halfway point. And there, suddenly, inexplicably, you see it—a person dressed up in a gorilla suit, standing smack in the middle of the basketball players, beating its hairy chest, looking right at you. Rewind,

*You can see this video at http://www.theinvisiblegorilla.com/videos.html.

and you see the whole impossible action. The gorilla strolls up to the players, turns toward the audience, thumps away, turns, and walks off slowly. Half the people who see this video fail to notice the gorilla.

Why? How could you fail to notice a monstrous ape amid ball-tossing college kids? It's because you are so deeply engaged in counting the number of passes that a gorilla is not enough to draw your attention away from the ball. You look right at the hairy beast and do not see it.

We've shown this video ourselves in dozens of lectures. We often ask people who do see the gorilla, "How many passes did you count?" The answer is usually wrong, or they admit to having not counted at all. Ironically, the better you perform the counting task, the less likely you are to notice the strolling gorilla. In other words, your focused attention ensures optimum performance in a given task but makes you blind to seemingly irrelevant data that may be more critical than the task at hand. Our own research shows that the brain suppresses distracters more strongly during a difficult task (when you are trying very hard to focus) than during an effortless task (when you are having an easy time). In everyday life, this means that even when you are focusing on accomplishing some critical job, you still need to remember to look up and around once in a while or you'll risk missing important facts and potential opportunities.

The Gorilla in Our Midst experiment raises an interesting question. Where are your eyes looking? Is the ball the only thing falling on your retina? Or is the gorilla's image also reaching your eyes but not registering with your brain? Eye tracking devices might help find an answer. An eye tracking device measures eye position under experimental and natural conditions. For instance, with a video camera pointed at the eyes, a computer program can find the pupils in the camera's image and detect how much they rotate from moment to moment. This allows scientists to know what the eyeball is looking at.

In 2006, Daniel Memmert showed, using eye tracking recordings, that many people do not notice the gorilla even when they are looking directly at it. Those who miss the gorilla spend as much time (around a second) looking at it as those who see it. This was an incredibly surprising result. Many neuroscientists had assumed that the gorilla was invisible because the basketball game drew the observers' eyes around the image, but away from the gorilla, at any given instant, as in overt misdirection. Memmert's results showed that they were wrong; it was really covert misdirection. The gorilla was invisible even when you looked right at it, because the basketball counting task drew your attention away from the gorilla. The study indicates that visual perception is more than photons entering your eyes and activating your brain. To truly see, you must pay attention.

Eye tracking has also been used to study attention and magic. In 2005, Gustav Kuhn and Benjamin Tatler, in the first study to correlate the perception of magic with a physiological measurement, employed an eye tracker to follow the eye movements of people watching a trick where a magician makes a cigarette "disappear" by dropping it in his lap. The researchers wondered: Do you miss the trick because you do not look at the right time? Or do you not attend no matter where your gaze falls? They found that the failure to notice the cigarette drop cannot be explained at the level of your retina. Detection rates were not influenced by blinks, saccadic eye movements, or the cigarette's distance from the center of the observer's vision at the time of the drop. The magician manipulates your attention rather than your gaze.

Inattentional blindness can get you into trouble in everyday life. How often have you been chatting away on a cell phone, only to find yourself bumping into another pedestrian? In 2009, psychologists at Western Washington University looked at four categories of college students walking across a main campus square. One set simply walked along minding their own business. A second walked in pairs, talking. A third listened to iPods as they walked. The fourth was gabbing on cell phones. In each instance, an outrageously costumed clown on a unicycle pedaled up to the students, circled them with comic abandon, and rode off.

Students who walked in pairs were most likely to see the clown. Those using iPods or walking alone were only slightly less attentive. But half of the students talking on cell phones entirely missed the clown on the unicycle. They also walked more slowly, weaving as they crossed the square. The researchers concluded that cell phone conversation leads to inattentional blindness and disrupts attention. It even disrupts walking.

ON MULTITASKING

Think you can text while driving? Listen to music while you pay your bills, send tweets, and monitor a football game on television? Write an e-mail, play solitaire, and check stock quotes while you have an argument with your spouse?

Think again. A decade of research clearly shows that multitasking—the ability to do several things at once, efficiently and well—is a myth. Your brain is not designed to attend to two or three things at a time. It is configured to respond to one thing at a time.

Research shows that you can't simultaneously give full attention to both the visual task of driving and the auditory task of listening, even if you use a hands-free device. In fact, people who talk on cell phones while driving a car have the same attentional focus as people who are legally drunk.* When you attend to the phone conversation you "turn down the volume" on the visual parts of your brain and vice versa.

Studies also show that people who are bombarded with several streams of electronic information do not pay attention, control their memory, or switch from one topic to another as well as those who complete one task at a time. Chronic multitaskers "are suckers for irrelevancy," says Stanford communications professor Clifford Nass. "Everything distracts them." They can't ignore things, can't remember as well, and have weaker self-control.

Another of our colleagues, Russ Poldrack at UCLA, has shown that people use the *striatum*, a brain region involved in learning new skills, when they are distracted and the *hippocampus*, a region involved in storing and recalling information, when they are not distracted. "We have to be aware that there is a cost to the way that our society is changing, that humans are not built to work this way," says Poldrack. "We're really built to focus. And when we force ourselves to multitask,

*The same problem does not happen when you talk to a passenger in your car because both of you will quiet down or stop talking when traffic gets heavy, it starts raining, or you need to make a quick lane change. Your passenger sees what you see whereas the person on the cell phone does not.

we're driving ourselves to perhaps be less efficient in the long run even though it sometimes feels like we're being more efficient."

Magicians know that multitasking is an urban legend and so they use a "divide and conquer" approach with attention: they split your attention so you cannot concentrate fully on any part of the stage at a given time. When your task list is pages long, you may feel tempted to do two or more things simultaneously. For instance, answering e-mail on your iPhone while attending a staff meeting. Chances are, you will do neither task well. For best performance, do one thing at a time.

Eric Mead, the mentalist whose knowledge of human nature never ceases to amaze, has joined the two of us at the Monterey Fish House in California, where we are bibbed and slurping down giant bowls of cioppino and glasses of Chianti. Susana asks Eric if he ever uses his training as a magician in everyday life. Without missing a beat, Eric closes his eyes and describes in detail the diners sitting all around us—how many are at each table, their genders and approximate ages, what they are having for dinner, even their conversations and apparent dispositions.

The couple on the left are celebrating a birthday. The family in the back attended a funeral earlier in the day. The ceremony was presumably for someone outside their immediate family (since they're here for dinner) but close enough to garner funereal attendance by the whole clan. The people behind Susana are in an unhappy marriage. To Steve's right a group of coworkers are celebrating someone's achievement; Eric's not yet sure what it is. The man over there is having a good time. That woman is in a bad mood. The birthday couple are making bedroom eyes at each other and do not want to be disturbed.

Eric says that he needs this kind of information when choosing subjects for his mentalism performances and he gathers it by practicing *situational awareness*—the deliberate perception of everything happening in the immediate space and time, the comprehension of its meaning, and the prediction of what may happen next. As we entered the restaurant, sat down at the table, ordered from the menu, and dug in, Eric casually cast his spotlight of attention onto all of the people around us, out of habit.

Eric never stops assessing his surroundings. You never know when you're going to need information for an impromptu display of magic, he says. By moving his attentional focus like a searchlight in the night sky, Eric has learned to avoid getting too absorbed by any individual aspect of what is happening around him and, for this reason, he says that he no longer experiences magic the same way most people do. He's not impervious to misdirection but he's resistant to it. Nor by his own admission is he any good at multitasking. The skill he describes involves serial attention.

We wondered how difficult it would be to learn situational awareness skills and attended a training course at the Marine Corps' Aviation Survival Training Center in Miramar, California. The navy teaches its aviators about situational awareness—how to optimize perception and cognition in demanding environmental and mental workload conditions. It does not matter if you are ordering from a menu while maintaining conversation or recovering from a flat spin in a fixed-wing jet, some optimal pattern of attentional scanning will maximize your success at whatever you are trying to achieve.

We experienced the challenge firsthand when we strapped in and flew a multimillion-dollar simulator of one of the largest helicopters in the U.S. military inventory, the CH-53 Super Stallion. Seated in the cockpit, we tried to allocate our attentional systems and scan our instruments while flying the huge beast. Our instructor, marine pilot Captain Vincent "Fredo" Bertucci, explained that your ability to scan your surroundings breaks down when your attention gets stuck in a rut. The world outside your windscreen beckons while your sensations give you the wrong information. Problems arise with the engines, with the ship you're landing on, with the load you're trying to lift with your chopper, with your communication systems inside and outside the aircraft. All of these events call for your attention, and will do so for too long if you're not careful. While your attention spotlights the one problem without scanning the other potential problems—for instance, you stare at a single broken gauge—you can unwittingly fly the helicopter into the drink.

Magicians use overt and covert misdirection to produce effects similar to these flight conditions. They split your attention and lead you to cognitive disaster. If we can reverse-engineer how magicians do it and apply those principles to developing methods to counteract attentional slips, we may be able to reduce the failures of attention that take place under conditions of high mental workload.

Two years after the Magic of Consciousness symposium in Las Vegas, we are in the quaint Pyrenees village of Benasque, Spain, attending an international conference on art and science. It is an eclectic group of experts who have come to explore the limits of human perception. Chefs are paired with scientists who study the sense of smell, architects are teamed with experts on human spatial perception, painters are linked up with visual neuroscientists, and the two of us are paired with one of Spain's premier young magic talents.

While we tackle the more academic aspects of overt and covert misdirection and their relationship to the brain's mechanisms of attention, Miguel Angel Gea cuts to the chase by performing tricks that dazzle the assembled cognoscenti, proving their grasp on reality is ever so frail.

Miguel Angel is a big young man with a long mane of brown hair cinched up in a ponytail. With his cargo pants and gauze shirt, he exudes casual good humor, which is not surprising, given that he was trained by Juan Tamariz himself. Miguel Angel is such a fun-loving soul that, despite his original intention to join us in Benasque for less than twenty-four hours, he ends up staying for four days—all due to the warm reception he receives from the conference participants and villagers. Our joint conference presentation begins at 9:00 p.m. and runs, by popular demand, until midnight, after which Miguel Angel repairs to the bars and restaurants of the village, regaling locals (who know him from Spanish television) with more tricks until the wee hours of the morning. He does this each night, ending the revelry only when he announces that he is completely exhausted and can no longer hold a coin or deck of cards.

Miguel Angel's love of life is profound. But so are his insights into human behavior. He uses the latest cognitive science literature as a lamppost for guiding the development of new tricks. For example, our colleague Dan Simons of "gorillas in our midst" fame designed another clever experiment that illustrates change blindness. In one version of the experiment, a proverbial absentminded professor is observed crossing a campus courtyard. A student walks up to the professor and says, "Excuse me, sir. Can you tell me where the gymnasium is?" He pulls out a campus map. "I don't know my way around."

The professor, who is happy to oblige, looks down at the map in joint attention with the student and begins to point the way. But just then two workmen carrying a large rectangular object—sometimes a door, sometimes a large painting—approach and endeavor to get by. "Excuse us. Excuse us, please, passing through," they say as they carry the object between the professor and the student. It takes but a couple of seconds, during which comes the switcheroo. The student—perhaps dressed in jeans and a red T-shirt, with dark hair—ducks behind the object and moves off. A second student who was crouching and moving behind the object—perhaps with blond hair and several inches shorter, dressed in slacks and a collared shirt—now stands up in his place. He is holding the map as he sidles up to the professor, who, amazingly, fails to recognize the change. Perhaps students are "homogeneous units" in his mind, but still you have to marvel at his change blindness. The experiment has been replicated many times, switching characteristics such as height, accent, and clothing of all kinds.

CHANGE BLINDNESS IN ACTION

When we first moved to our current institute, Susana needed an additional laboratory room in which to conduct perceptual experiments. Her department chair graciously offered the drafting room, as long as Susana didn't mind sharing her space with a lot of bulky equipment—a slant table, large cabinet with flat drawers to store large drawings, huge paper cutters, and the like. Susana gratefully moved in. She then went to the individual lab heads and asked if they wouldn't mind removing any drawings they had stored in the cabinet, since the piece of furniture simply took up too much space. Each person graciously agreed to help and Susana got rid of the cabinet. Then, on a different floor of the research building, Susana found another shared equipment room with some counter space available. She moved the paper cutters and other equipment out of her new lab room. Over the course of a few weeks, the drafting room became a drafting room in name only, as it had been completely transformed into her lab. So Susana called the facilities staff and asked them to change the sign on the door from "Neurobiology Drafting Room" to "Laboratory of Susana Martinez-Conde."

Slowly but surely, Susana had transformed her perceptual testing lab space from a single corner of a shared drafting room to her own complete unshared lab space, all by employing the principles of change blindness. The department chair still shakes his head when he's reminded of it, but he's never asked for the room back because the lab space is very productive and its projects have earned grant money to support the research in the room.

Miguel Angel figures that if you fail to notice two very different people swapping places, then you can miss just about anything. Certainly you can mistake one card for another. One afternoon at the conference he demonstrates how. Dressed in his usual casual attire, Miguel Angel calls for a volunteer from the audience. Once she is onstage, he asks her to pick a card from a deck. It is the eight of clubs. He shuffles it back into the deck. "I like to pull your card from my pocket," says Miguel Angel as he magically pulls the eight of clubs from his right hip pocket. Applause.

He looks at the volunteer. "Did you like that trick? Yes? There are tricks that I don't like." He raises his empty hand toward her and reaches into her hair, and, as he pulls away, the eight of clubs is back in the palm of his hand. "Other magicians like to pull cards from people's hair. But I don't like that trick myself," says Miguel Angel. Snickers emanate from the audience.

Next, Miguel Angel slides the eight of clubs back into the deck and places the deck on a table, holding a few cards out. He then rubs those few cards between the thumb and fingertips of his right hand.

"Other magicians prefer to make coins appear," he says, as a large coin slides out from between the rubbed cards into his left palm. The crowd responds with oohs and aahs.

The volunteer shakes her head in disbelief. He looks at her as he deposits the remaining cards on the table—which is now a full deck that obviously includes the eight of clubs—leaving only the coin in his left hand. He tosses it into his right palm. "But me, no. I don't like tricks in which cards are taken from your hair, or even tricks with coins," he says as he tosses the coin back again—but this time it disappears.

"No," says Miguel Angel, "I prefer tricks with a single card in my pocket." He dips his empty right hand into his pocket and pulls out a card with its back to the audience. "And this single card is your card," he says as he rotates it forward to miraculously reveal the eight of clubs. Wild applause.*

Miguel Angel has a sly smile on his face. He turns to the audience. "Would you like to know how that trick worked?" We shout a resounding "Yes!" He stands there for a second, as if contemplating his next move. He seems suddenly awkward. "It's a little difficult for a magician to reveal a trick," he says sheepishly. The audience laughs, sympathetically, as Miguel Angel reaches a decision. Flinging his arms over his head, he pronounces, "For science!" and launches into an explanation that fascinates scientists and artists alike. He reveals that the trick starts before the show when he picks two similar-looking cards, in this case the eight of clubs and the eight of spades. He places the club above the spade and puts the deck in his pocket for safekeeping until the show.

When Miguel Angel asks the volunteer to pick a card, he uses a *force* so that she chooses the eight of clubs without realizing it. *Forcing*

*Miguel Angel Gea's routine is an expanded version of a trick by Francis Carlyle. The card swapping technique was already used by Johann Nepomuk Hofzinser, a Viennese magical genius from the mid-nineteenth century.

refers to a number of methods used by magicians to make you think you are freely choosing a card whereas they know in advance exactly what card you will take. We'll talk about this in more depth in the next chapter.

When the volunteer puts the eight of clubs back into the deck, it is not randomly inserted. Miguel Angel again forces her to place it where he wants it—directly above the eight of spades. His subsequent moves are stock sleight of hand. He "shuffles" the deck so that the two black eights are on top. He palms them and drops both in his pocket. When he says "I like to pull your card from my pocket," he reaches in for the eight of clubs and leaves the spade behind. (You can probably see where this is going.) He then works it so he "pulls" the eight of clubs from the volunteer's hair and he uses the coin routine as distraction from his main goal, which is change blindness.

When all the cards are on the table, you assume the eight of clubs is safely somewhere in the pile. That's when Miguel Angel reaches into his pocket and removes the eight of spades. He finishes his routine by saying, "Now, I prefer tricks with a single card in my pocket," and he flips the eight of spades over. But you and everyone else are so eye-rollingly amazed, so completely enthralled by the fact that Miguel Angel has impossibly produced the eight of clubs from his pocket, when it's supposed to be on the table, that you fail to detect that it's not the eight of clubs at all. It's a spade. Even the volunteer, less than three feet away, looks at the spade but fails to see it's the wrong card. Miguel Angel manages to fool one hundred leading scientists and artists with a classic example of change blindness.*

END OF SPOILER ALERT

Change blindness studies show that you will not notice dramatic changes in a visual scene if they occur during a transient interruption—such as a magician reaching behind the ear of a spectator, or two workmen carrying a door between you and the person you are talking to—even when you are looking right at the changes.

*For more on Miguel Angel, see http://www.miguelangelgea.com and http://www.sleightsofmind/media/miguelangelgea/.

Change blindness is also common during cuts or pans in movies. A wineglass may be empty in one scene and full the next scene. Chances are you'll miss it.

Slow or gradual changes are also very difficult to see, especially if we are not focusing our attention on the changing object. This has been compellingly demonstrated by Simons: whole buildings, boats, people, and other highly salient objects may appear and disappear unnoticed, right in front of our eyes, if they do so slowly enough. It is tempting to speculate on how many things in our lives may slowly change without our awareness. The small aches, pains, and debilities that colonize our bodies as we age would be intolerable if suddenly imposed on a healthy twenty-year-old, but as we gradually grow older these changes creep in for the most part undetected. Other aspects of our lives, jobs, and relationships may similarly change, worsening or improving in a very gradual and thus unnoticed fashion.

The Greek philosopher Epicurus knew that we tend to adapt to and thus ignore gradual improvements in our lives. He wrote: "Do not spoil what you have by desiring what you have not; but remember that what you now have was once among the things you only hoped for." It's sage advice, provided your pocket isn't being picked while you're distracted by your gratitude.

6

THE VENTRILOQUIST'S SECRET

Multisensory Illusions

The first thing we notice on arriving at the 24th World Championship of Magic in Beijing is that the massive building where it's being held—the China National Convention Center—is all smoke and mirrors. It's not that there's something wrong with the air handling system, or that a magician used too much dry ice during his act. Rather, the edifice was constructed with mirrored panes of glass that, in the high heat of July 2009, trap huge shrouds of urban smog. Outside, it's worse. The entire city of Beijing is blanketed in smog so thick that you feel as if everybody and everything around you is an apparition emerging from behind a magician's smoke machine.

Plenty of venues award magicians for their skills, but the World Championship of Magic is the preeminent international event. Held every three years in a different country by the Fédération Internationale des Sociétés Magiques (FISM), this competition is informally known as the Magic Olympics. The weeklong contest is where magic stars are created. Winning a grand prize here is akin to receiving an Oscar; it's a guarantee of steady work for years to come. Many talented young magicians such as Lance Burton have gone from obscurity to world-class fame at the Magic Olympics. We're here to see it happen with our own eyes, and it's quite a scene. Twenty-five hundred amateur and professional magicians, purveyors of magical paraphernalia, and curious onlookers from sixty-six countries amble through the main lobby and corridors on their way to huge festooned halls where the ceremonies and competitions are being held. Their attire ranges from standard street clothes to wizard's robes and everything in between.

Some people attend lectures given by famous magicians on topics such as "From chaos to order: different methods to secretly arrange the cards into special order," "Japanese style on how to study magic," "How to present the same trick three different ways," and "Boldness and magic, or the art of having real nerve." Others roam between booths selling rope and card tricks, fake appendages of all types, magic trick books, special order card decks, stage gimmicks . . . everything a magician could ever covet.

One hundred performers are competing for the grand prize in the two main categories of stage magic and close-up magic. Stage performances are judged on manipulation, general magic, stage illusions, mental magic, and comedy magic. Close-up performers are rated for card magic, parlor magic, and micromagic (magic tricks done on a very small scale, such as small coin tricks or tricks with toothpicks).

Two days into the event we are thrilled to spot Max Maven, one of the world's greatest living mentalists, sitting in front of the overflow screen in the main hall outside the competition rooms. Maven is legendary for his ability to read minds. Onstage he assumes a sinister Svengali look: thick black eyebrows arched high with disdain, a clipped Fu Manchu mustache, and a meticulously shaved widow's peak. He has an extremely high, heart-shaped forehead, pointy ears, a deep baritone voice, and salt-and-pepper hair pulled back into a taut knot from which a long braid hangs down his back. To complete his look, Maven wears black double-breasted suits, black shirts, and bold silver bracelets and rings.

But today Maven is in his street clothes—black T-shirt, black pants and boots. His samurai hair is loosely braided and almost reaches the floor. It is late afternoon and beams of sunlight are shining into the convention center's halls like pillars of gold in a cathedral. A large sign—RM 319, RESTAURANT AND MAGIC SALON—hangs on the wall.

Maven is relaxing and watching a twenty-foot-high movie screen where a young magician from Sweden is producing card after card after card from an empty outstretched hand. Our son Iago has fallen asleep in his stroller, so we roll over to Maven. We've got a question ready for him: Does he know any multisensory tricks? That is, can he tell us about tricks that rely on interactions among the different senses such as vision, hearing, and touch? Maven is pleased with the query and responds by telling us a classic joke used by generations of magicians to entertain friends and family. It's the Dinner Roll trick.

Here's his description: To begin, the magician is seated at a dinner table covered with a cloth. He makes sure that you are in front of him, unable to see his movements behind and below the cloth. He says something corny like "You know it's rude to play with your food. But I wonder if this soft dinner roll will bounce?" He holds the roll in his hand and flings it to the floor. You hear it bounce with a loud *thunk* and then fly back up into the air, where he catches it.

The secret behind this very convincing illusion is simple. The magician sits across the table from you, turned away from the normal eating position. The magician's hand makes the motion to fling the roll to the floor. As soon as his hand and lower arm are out of your view, he turns his hand over, palm up. Using his fingers and wrist, he launches the roll back up into the air, making sure not to move his upper or lower arm. All the action is in his fingers and wrist—and in his foot. Before the roll reappears in midair, the magician taps his foot. You hear the *thunk* at the same time the roll would have hit the floor.

But what makes the dinner roll trick really interesting is a twist performed by the magician Jay Marshall. He put an extra delay between the sound of the roll hitting the floor and its bounce back up. It's as if the roll dropped down *below* the floor before it hit and bounced back up. This maneuver heightens the illusion, and nobody notices the discrepancy.

END OF SPOILER ALERT

In movies, technicians called Foley artists artificially exaggerate everyday sounds to make them more realistic. For example, they might re-create the sound of walking in the mud by rhythmically squeezing a wet newspaper in time with the screen actor's footsteps. A recent study showed that listeners deem such modified sounds more realistic than recordings of the actual event more than 70 percent of the time. Susana witnessed this when she joined a gym to practice tae kwon do (a Korean martial art) at the age of fifteen. On her first day she was surprised to find out that unlike punches in action movies a real-life punch doesn't make much noise.

Another multisensory trick popular with dinner table magicians involves a disappearing saltshaker. Again seated across the table from you, the magician puts a quarter on the table and says, "Would you like to see me make this coin go through the table?" Of course you would. The magician explains that he'll need a bit of help moving the coin. He takes the saltshaker, wraps it snugly in a dinner napkin, and taps the coin. *Tap tap.* He moves the napkin-clad shaker back toward his body. Nothing happens. The coin is still there. He repeats the *tap tap* and movement of the shaker. The coin has not moved. He does this a third time, saying, "Oh, my, this is difficult," and leaves the shaker on top of the coin. Then he takes his hand and *wham,* he slams the saltshaker right through the table. At least that is what it looks like. The saltshaker is gone. The napkin is flat and the quarter is still on the table.

SPOILER ALERT! THE FOLLOWING SECTION DESCRIBES MAGIC SECRETS AND THEIR BRAIN MECHANISMS!

This trick, too, is simple. The second time the magician pulls the saltshaker back toward his body, he deftly takes it to the edge of the table and drops it into his lap. Because the napkin retains the shape of the shaker, you assume it is still in his hand, within the napkin shroud. Meanwhile, the magician uses his free hand to move the actual saltshaker under the tabletop to the position directly below the coin. He makes a third *tap tap tap* motion with the napkin, but this time the sound actually comes from below. When the magician slams the empty shaker-shaped napkin flat, your sense of vision and hearing together create the perception that the saltshaker has passed through the table. It's a profoundly convincing combination.

END OF SPOILER ALERT

These two tricks reveal a fundamental property of your brain: your propensity to integrate information from multiple senses as you interact with the world. When you simultaneously see a bright light and hear a loud sound, your brain figures they are related. Recall that illusions occur when the physical reality does not match the perception. If you see cymbals banged together and hear the resultant crash, it's not an illusion. But if you're in Boston for the Fourth of July celebration and you see the Boston Pops cymbals banged and hear only the howitzers firing during Sousa's crescendo, it's an illusion.

The fact that your brain combines sights and sounds into single perceptions seems patently obvious, but for neuroscientists the phenomenon is surprisingly complicated. From Aristotle on, researchers have tended to study senses—sight, hearing, touch, smell, taste, balance, self-motion, and feelings from the body—in isolation. Magicians, on the other hand, have learned to manipulate perception by understanding when and where the senses don't mix accurately.

Are senses really separate? When you encounter the world, your experience is not disjointed. When you perceive a barking dog, you don't feel you are seeing it with one channel of your brain and hearing it with another. In general, when combinations of sounds, smells, tastes, lights, and touches occur simultaneously, you perceive a coherent multisensory world.

Senses not only interact, they enhance one another. For example, what a food sounds like can determine how it tastes. Potato chips are yummier if they make more noise when you bite into them. Bacon and egg ice cream (sorry, the experimenters are British) tastes more bacony if you hear the sound of bacon sizzling in a pan, more eggy if you hear chickens clucking in a farmyard. Oysters taste better when you listen to seagulls and crashing ocean waves.

The same goes for skin and sound. When you say a word that begins with the letter *p*, *t*, or *k*, you produce a puff of air that is sensed by mechanoreceptors in human skin. The puff of air helps you and others perceive the sounds correctly. This rather amazing fact was revealed in a recent series of experiments. If you were a participant, you would sit in a chair while researchers delivered tiny puffs of air to your ankle and played the sounds *pa* and *ta*. You would hear *pa* and *ta*. But when they played *pa* and *ta* without the puff of air, you would more likely hear *ba* and *da*.*

Your eyes can fool your ears. Check out the McGurk effect.† In this auditory-visual illusion you will see a film clip of a man saying "da da da." But if you close your eyes, you will hear him saying "ba ba ba." Then if you mute the sound and just watch his lips, you will clearly see that he is saying "ga ga ga." The effect is amazing. It happens because your brain does its best to reconcile mismatching information whenever it can. Sometimes your brain's best is not good enough to be accurate. But then again, it is very unlikely that you will ever see "ga ga ga" and simultaneously hear "ba ba ba" in nature. The reason these effects work is that your brain takes shortcuts to make likely interpretations of perceptions occur faster. Thus, although the resultant perception may not be accurate (it's an illusion because the perception does not match the physical reality), the illusion is *accurate enough* and has helped humans to survive by saving brain processing time and effort as, for instance, your ancestors listened for leopards prowling in the nearby bushes.

Your ears can also fool your eyes. If you look at a single flash of

*Actually you do mishear people all the time. You figure out the meaning from the context. You may hear "bog" instead of "dog" but the sentence "the boy petted the bog" makes no sense—so you think dog. Moreover, even though you can hear sounds in the absence of airflow, puffs of air might make it easier to distinguish between two words such as "tall" and "doll" when there is a lot of ambient noise.

†See http://sleightsofmind.com/media/McGurk. This is sometimes called the McGurk-MacDonald effect. It was first described in 1976 by Harry McGurk and John MacDonald in "Hearing Lips and Seeing Voices," *Nature* 264: 746–48.

light while hearing multiple beeps, you may see multiple flashes. In the same vein, what you hear influences what you feel. In the parchment skin illusion, you rub your palms together while listening to different sounds. Higher frequencies will make you feel as if your hands are rough. Lower frequencies give you the impression of your hands being smooth, although nothing about them has changed.

How you feel the world can actually change how you see it and vice versa. Remember the waterfall illusion from chapter 1? If you stare at the downward motion of a waterfall for some period of time, adjacent stationary objects such as rocks appear to drift upward. But if you feel an up or down sweep on your fingertip as you watch the waterfall, the perceived direction of water flow switches. Touch alters vision.

And then there is the rubber hand illusion, which you can try at home. First you need to buy one of those creepy rubber hands from a Halloween store. Let's assume it's a right hand. Sit at a table and place the hand on the table where you can see it while putting your own right hand in your lap, out of sight. Ask a friend to take two soft paintbrushes and simultaneously stroke your real hand and the rubber hand with the same rhythm. If you are like many people, you will soon feel that the dummy hand is your own. If your friend smashes the rubber hand with a hammer, you may scream *Ouch!** With the proper equipment, including a virtual reality headset, you can even induce an entire out-of-body experience based on this illusion.†

A surprising number of people experience unusual multisensory perceptions because of the way their brains are wired. One sensation, such as music, triggers another type of sensation, such as taste. Senses are cross-activated. For example, some people perceive letters or numbers as having color. For one person *A* is always red, *B* is always turquoise. For another person, 7 is always yellow, 4 is always orange. Days of the week can possess personalities: Tuesday is sad, Wednesday is happy. These associations are idiosyncratic and automatic, lasting a lifetime. The phenomenon is called *synesthesia.*

Neuroscientists have identified at least fifty-four varieties of syn-

*See http://sleightsofmind.com/media/rubberhand.
†See http://sleightsofmind.com/media/out-of-body.

esthesia, including some that are quite common. People with auditory synesthesia hear sounds such as tapping, beeping, or whirring when they see things move or flash. This trait was discovered accidentally when a student participating in a study of visual motion reported hearing sounds when observing a scene similar to the opening of *Star Wars*, when the stars fly out at you, but in this case there was no sound track. Researchers soon identified many other students with the same cross-sensory perceptions. It seems that some people have an enhanced sound track to life, which makes sense when you consider that in the natural world many moving things (say, a bee) make sounds when they move (buzz).

In time-space synesthesia, a visual experience can be triggered by thinking about time. Like Kurt Vonnegut's Tralfamadorians in *Slaughterhouse-Five*, some people can literally see time. For example, some say they view a year's time like a circular track with them standing in the middle. They can see the days and months unfolding all at once.

In mirror touch synesthesia, which is rare, people experience sensations of touch on their own bodies when they see other people being touched. They sense a slap on their shoulder when they see another person get slapped on the shoulder. Same goes for a kiss.

Synesthesia runs in families, suggesting a genetic origin to the condition. It is common for family members to experience different types of synesthesia and for the trait to skip generations. Research shows that synesthesia is caused by increased cross talk between various brain regions as well as extra connective pathways linking them.

As for the rest of us, synesthesia offers insights into our everyday perceptions. All of us have our sensory wires crossed to some extent, if only to process multisensory inputs. Look at the two shapes on the following page. Which one would you call *bouba* and which *kiki*? If you are like the vast majority of people tested from a wide variety of language groups, you will identify the rounded shape as *bouba*, maybe because your mouth makes a more rounded shape to produce the sound. When you articulate *kiki*, your mouth is more angular with the harder sound of *k*. Such synesthesia-like mappings may be the neurological basis of how sounds are mapped to objects and actions in the world.

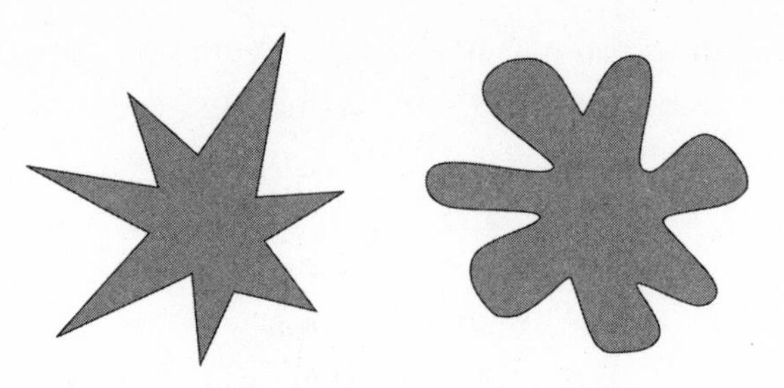

The *bouba kiki* effect was first described in 1929 by Wolfgang Köhler. The vast majority of people identify *kiki* with the angular figure and *bouba* with the rounded one.

Magicians intuitively know that your senses interact. They know they can fool you by the sound of the dinner roll hitting the floor and the sight of it bouncing back up into the air. When the magician pretends to throw the bun on the floor, you hear a *thunk* and perceive the act as having happened. Your brain integrates the sight and sound of the bun into a single perception: it landed and bounced. It is a multisensory illusion. By adding a delay, magicians discovered how to make the roll into what scientists call a *superstimulus*. They are live-action Foley artists. Similarly, when you see the saltshaker-shaped napkin tap the table and hear the sound of tapping, your brain integrates the sight and sound, leading to an auditory-visual illusion.

A superstimulus is a supersalient object or event that evokes a stronger neural and behavioral response than the normal stimulus for which the response evolved in the first place. It's supersized fries when you're hungry. It's an ice-cold pint of beer when you're thirsty. It's the extreme curviness and abnormally large breasts of the prehistoric Venus of Willendorf. It's mascara and lip gloss on a sexy female face (remember from chapter 3 that increasing the contrast of eyes and lips produces the illusion of making a face look more feminine). Superstimuli invite attentional focus. Jay Marshall realized that the timing of the roll hitting the floor was the key to the illusion. By increasing the delay enough to engage interest, but not enough to generate an incongruity, he made the bounce of the dinner roll seem more magical.

Multisensory integration is an ongoing and dynamic property of your brain that occurs outside conscious awareness. So, where in

The Venus of Willendorf is a female superstimulus from more than twenty thousand years ago.

your brain does the cacophony of sensory information come together? Your senses are separate, in that your eyes, ears, nose, skin, and tongue are located on different parts of your body. But your experience is coherent, integrated, and usually unambiguous.

Part of the answer is that you possess multisensory neurons. Just as you have neurons that specialize exclusively in vision, hearing, or touch, you have neurons that fire in response to simultaneously occurring sights and sounds, touches and sounds, touches and sight, and so on for all aspects of sensory processing (including balance and pain and the location of your body in space).

Multisensory neurons are found throughout your cortex, even in areas thought to specialize in a single sense. For example, several brain regions traditionally thought to be visual areas have multisensory neurons that fire in response to sounds and/or touch. And a midbrain region called the *superior colliculus* is densely packed with multisensory neurons that map out your brain's responses to all of these senses. Superior colliculus neurons extract clues from multiple sources, including multisensory neurons in your higher cortex, and help you orient your head and body to what's important in the world at any given moment.

Have you ever driven a cat crazy with a laser pointer? The cat will chase the little red spot up a wall, under a rug, behind a couch, or wherever you point the thing. It's great fun for feline and human alike. Researchers recently borrowed this game to run a multisensory experiment. Cats were trained to look straight ahead and then approach a very low intensity light, which was a very demanding task. But when the scientists added a brief, low-intensity burst of noise from the same location as the light, the cats performed brilliantly. When the researchers added a soft sound from another location, the cats failed miserably.

Now imagine a cat hunting mice in the dark. The rodents make soft skritching noises while the cat's sensory whiskers sweep the environment. For a cat, whiskers are better than eyes. By combining sound with whisker motion, the cat triumphs. The lesson for mice: keep silent.

But a deeper question remains. While multisensory neurons can combine inputs from your different senses, they are still widely located throughout your brain. It's not plausible that every multisensory neuron is directly wired to every other multisensory neuron. So how do they fire in concert? Objects have different features such as color, shape, sound, or smell. How does your brain figure out which features belong to the same object? How are unified conscious experiences bound in your brain? How does your brain connect the sight and sound of Marshall's bun?

Called the *binding problem*, the question has many proposed solutions but no definitive answer to date. People might possess a single neuron for each possible combination of features, but that is unlikely given the sheer number of combinations. One solution, called *feature integration theory*, posits that binding is accomplished by an act of selective attention. It occurs within your spotlight of attention when your brain's circuits combine different types of features of an object such as its color and brightness or shape and sound. The integration of two or more features speeds up the detection process and helps you to quickly adjust your attention to focus on the task you are performing. While neuroscientists have not settled on a solution to the binding problem, magicians merrily exploit the fact that attention-grabbing information from one sensory system leads to enhancement of attention in another. Thus a magician's rapid-fire patter serves to increase how intently you stare at the actions he wants you to look at. The tapping of the saltshaker under the tabletop right before it appears to sink through the table forces your brain to pay attention and visualize the false event.

Next time you log on to the Internet, go to YouTube and type in "Terry Fator." You won't be disappointed. Fator won first prize in the 2007 *America's Got Talent* competition with his ventriloquism act. His puppets impersonate famous singers—Roy Orbison, Elvis Presley, Marvin Gaye, and many more—while Fator's lips never seem to move. The judges swooned. The audience screamed with delight. The Mirage Hotel in Las Vegas saw a good thing. Fator now has a five-year multimillion-dollar contract and his own theater to bring ventriloquism, an ossified art, into the twenty-first century.

Ventriloquism is the feat of shifting sound toward a visual target. It is a classic multisensory illusion with deep historical roots. In many preagricultural societies, shamans used ventriloquism to speak with the spirit world. Inuits would descend into a netherland full of growly voices and appear to use a harpoon in battle. They "emerged" covered in blood (thanks to a bladder of blood stashed under their parkas) to reveal truth and wisdom. In Greece, at the Temple of Apollo at Delphi, "belly speakers"—"ventriloquism" means "speaking from the stomach"—gave voice to divine revelations and prophecy emanating from the dead. When you think about the world before the invention of recorded sound, you can appreciate the wonderment elicited by ventriloquists. Today we are accustomed to sounds coming at us from all directions, in elevators, shopping malls, restaurants, and so forth. But before Victrolas and, much later, iPods, a voice coming from above the ceiling or from beneath the floor (a favorite ventriloquist trick) could be terrifying. It was black magic.

During the Enlightenment, ventriloquism lost its reputation as black magic when magicians stepped forward to demonstrate the art of "throwing one's voice" and essentially demystified it. They described it for what it is: a multisensory illusion, one that takes an enormous amount of practice to make convincing.

Try saying "big love" without moving your lips. Or "mama papa." When you look at Fator's mouth, you'll notice that his lips scarcely move. His throat moves, but he hides it behind a goatee and a microphone. Like all ventriloquists, Fator employs a set of acoustic approximations and articulatory tricks. Sounds made with the lips—*p*, *b*, and *m*—are acoustically similar to sounds made by the tongue on the soft palate—*k*, *g*, and (soft) *ng*. He can substitute the latter for the former. By forcing air through his slightly parted mouth, Fator can make the sounds *f* and *v* without using his lips. All other sounds in English can be made with duplications inside the mouth.

In the early twentieth century, ventriloquists such as Edgar Bergen (and his dummy partner Charlie McCarthy) were enormously popular. Bergen danced with his dummies, made silly jokes, and brought vivid characters like Mortimer Snerd to life. But when another source of multisensory illusion—talking pictures—arrived, ventriloquist acts like Bergen's were doomed, displaced by the silver screen. For sheer entertainment there was no competition.

Next time you go to a movie theater, consider the fact that films are a form of ventriloquism in that speech is not coming from the actors' lips. Sound is being piped into speakers far removed from their actions. Your brain creates the illusion of actors talking to one another, thanks to your multisensory brain. Moreover, images appear to be stable when in fact they are flickering. The steady appearance of a flickering light source—such as a fluorescent light, display screen, movie, or television set—is known as *flicker fusion.* It occurs when the rate of flicker is higher than a critical threshold, which for motion pictures is 24 frames per second.

Flicker fusion is thought to occur due to a process called *persistence of vision.* The concept was first presented to the Royal Society of London in 1824 by Peter Mark Roget (who also wrote the famous thesaurus) as the ability of your retina to retain an image of an object for between one-twentieth and one-fifth of a second after it is removed from your field of vision. Johnny Thompson exploited this fact in his red dress trick.

Max Wertheimer, the founder of the famous Gestalt school of psychology mentioned in chapter 2, and Hugo Munsterberg discovered a second principle—the *phi phenomenon* or stroboscopic effect, which is closely related to flicker fusion. You can perceptually bridge the temporal gap between two consecutive displays so that you perceive a series of static images in a continuous movement. Add this visual illusion to a nearby sound source and your brain does the rest: you are seamlessly transported to wondrous fictional worlds (unless of course you are watching a poorly dubbed foreign film!). The interconnection between our senses also plays a role in magic tricks involving memory, which is the subject of the next chapter. Consider this story.

As a reporter in the 1920s Soviet Union, Solomon Sherashevsky was able to remember names, dates, directions, sources, and other newshound essentials—without ever writing anything down. His editor thought Sherashevsky was being lazy in staff meetings, since he didn't take notes of his assignments, and one day he asked the reporter to repeat every word of what had been said at that morning's briefing. Sherashevsky did so, flawlessly—somewhat amazed, we are told, that his talent was considered unusual. The dumbstruck editor sent Sherashevsky to the laboratory of Russian psychologist Aleksandr Romanovich Luria "to have some studies done on his memory." In the years that followed, Luria studied "the man with the vast memory," noting that his talent stemmed from a form of synesthesia. Sherashevsky saw vivid images—such as splashes of color or puffs of smoke—with every word, number, and syllable. Whenever he wanted to recall numbers, syllables, words, or events, he would conjure up the combinations of images in his mind's eye and report what he saw. In this way, he could remember almost everything he encountered. As we've seen, magicians, even if they don't have synesthesia, can capitalize on the mingling of the senses.

7

THE INDIAN ROPE TRICK

Memory Illusions

The 1890s are remembered as an age of exuberant invention, when steam age engineers developed early precursors of the airplane, automobile, and cinema. Wilhelm Roentgen identified X-rays, Marie and Pierre Curie discovered radioactivity, and William James described the principles of psychology. Readers were enraptured by Sherlock Holmes, Dracula, and Rudyard Kipling's *The Jungle Book*. But for millions of people immersed in contemporary spiritualism—replete with séances, psychic lights, voices of the dead, and dark secrets from the Orient—the best new thing may have been a breathtaking magic act called the Indian rope trick.

On August 8, 1890, the *Chicago Tribune* carried the first officially recorded account of the trick. Two Yale graduates, an artist and a photographer, were traveling in India when they saw a street fakir pull a ball of gray twine from under his knee, hold the loose end in his teeth, and toss the ball toward the sky. The twine unrolled until the other end was out of sight. A small boy, "about six years old," then climbed the twine. When the lad was thirty or forty feet in the air, he vanished. *Kapoof.* This happened outside in daylight; no hidden wires or supporting gizmos could be concealed from view. The artist sketched the event. The photographer took snapshots. But when the photos were developed, they revealed no twine, no boy. There was only the fakir seated on the ground. The anonymous author proffered an explanation: the fakir had mass hypnotized the entire crowd, but he could not hypnotize the camera.

According to Teller, who wrote about the trick several years ago, the story's genius is that it allowed many readers to wallow in Oriental mystery while maintaining the pose of modernity. Hypnotism was to the Victorians what energy is to the New Age: a catchall explanation for crackpot beliefs. By describing a thrilling, gravity-defying miracle, then discrediting it as the result of hypnotism—something equally cryptic but with a Western, scientific ring—the *Tribune* allowed its readers to have their mystery and debunk it too.

Four months after the article appeared, the editor of a British weekly wrote to the *Tribune* asking to speak with one of the Yale graduates. He received an apologetic note from the author of the article: "I am led to believe that the little story attracted more attention than I dreamed it could, and that many accepted it as perfectly true. I am sorry that anyone should have been deluded."

In other words, it was a hoax. The trick is impossible. It does not exist nor has it ever existed. Historians of magic say it is fitting that the author of the article was John Elbert Wilke, a gifted liar who later became the first director of the United States Secret Service, famous for his skulduggery and Machiavellian scheming. He wrote the story simply to increase the newspaper's circulation. Wilke then printed a retraction, noting that the story had been "written for the purpose of presenting a theory in an entertaining form." The byline of the retraction was Fred S. Ellmore (as in sell more papers).

But Wilke's retraction came too late. The story had already gone viral. Long before the Internet made the spread of information instantaneous, news of the Indian rope trick flashed worldwide—it just took months rather than minutes. The story was picked up by newspapers throughout the United States and Europe, was translated into nearly every European language, and also reached India, where it was met with surprise. What rope trick?

For the next fifty years, many hundreds if not thousands of people gave eyewitness accounts of having seen the Indian rope trick. In 1904, a young British gentleman, deemed trustworthy by virtue of his high breeding, told the Society for Psychical Research that he had seen the trick a few years earlier. After lengthy questioning, the society dismissed his testimony as illustrating "once more the unreliability of memory." But reports continued to pop up, with embellishments: After the boy disappears into thin air, the fakir calls for him to return. Hearing no reply, the fakir grabs a knife, climbs the rope, and also disappears. Shouts ensue. Then pieces of the boy—leg, arm, torso, head—fall to the ground. The fakir climbs down and puts the pieces of the boy into a basket; after an incantation, the boy jumps out, whole and smiling. The fakir is covered in blood.

As the legend of the rope trick grew, so did its pedigree. Historians traced it to the ancient world with antecedents in Australia, Siberia, Germany, and China. Indian scholars referenced rope-climbing metaphors from the eighth century. Marco Polo was said to have encountered the trick.

Magicians stepped up to debunk the trick, which they knew to be impossible, and offered a reward to anyone who could actually perform it. But every time they managed to discredit a sighting—such as proving that the "rope" in question was really a pole—more first-person accounts poured in. Like the Loch Ness monster, Bigfoot, or UFOs, the Indian rope trick retained credibility despite the lunacy of it all.

If you've never heard of the legend, it's likely because its popularity peaked in the years just before the Second World War. Modern magicians occasionally try to mount a version of the trick but routinely fail to deliver. It was not firmly and decisively revealed as a hoax until 2005 when Peter Lamont, a research fellow at Edinburgh University, published the full story in his book *The Rise and Fall of the Indian Rope Trick.* Lamont explains that the trick is a classic example of how memory illusions take root within the human mind.

Were eyewitnesses lying? A lasting human foible, Lamont says, is that people will believe hoaxes and rumors to be true despite all evidence to the contrary, including denials by their originators, if assertions of truth are repeated often enough. In this regard, the Indian rope trick shares features with modern political "controversies," such as the claim that Iraq possessed weapons of mass destruction, that Barack Obama was not born in the United States, or that astronauts never set foot on the moon. Another persistent human foible is the exaggeration effect. According to Lamont, the longer the period between when the trick was seen and when it was reported, the more impressive the account of it. In other words, people tend to confabulate over time. Indian street magic includes acts in which children climb poles, hide in baskets, and appear to be mutilated—all potential sources of confusion in memory formation. The true secret, Lamont concludes, is the way supple human memory combines events seen with legends only heard. We reshape our memories with each retelling of them, which means that along with your willingness to be misdirected, your memory is an easy target for magicians to exploit in countless tricks of their trade.

Johnny Thompson—the Great Tomsoni of the red dress trick—is happy to demonstrate how he manipulates memory. He has the perfect audience: the hundred or so scientists at the Magic of Consciousness symposium. They are trained observers. Can he fool them?

Johnny asks the scientists if they believe in mind reading or psychic or paranormal abilities. He calls up a volunteer, whose name is Dan, and asks again, "Do you believe in mind reading?"

"No."

"Neither do I. I'm a faker, fraud, phony, and cheat." But then, says Johnny, "nobody's perfect." Only he *is* perfect—at fakery. The trick, he says, is based on psychology, behavior patterns, and "closing the doors" to all rational explanations for what we are about to see.

Johnny takes out his wallet and removes a $100 bill. He also takes out a small envelope and asks Dan to examine but not open it. The envelope, he says, contains a prediction. The money and envelope go back into the wallet, which then goes into Dan's breast pocket.*

Next, Johnny pulls out a "perfectly ordinary deck of cards,"

*See http://sleightsofmind.com/media/magicsymposium/JohnnyThompson.

shuffles them, and asks Dan to cut the deck. From our angle, seated behind him, there is no apparent trickery going on. As far as we can tell, he doesn't put anything into his pockets or pull anything out. Then Johnny explains the challenge: there are fifty-two cards in the deck, and only one matches the card sealed in the envelope. All Dan has to do to win is pick one of the fifty-one cards that does *not* match. If he picks the card that matches, he loses.

After the cards are cut, Johnny asks Dan to turn them over slowly, one at a time, and stop whenever he "feels" a precognition that his chosen card will not match the one inside the envelope in his breast pocket. Dan stops at the nine of clubs. Johnny teases him. Is he sure of his choice? Doesn't he want to go one more card? Dan says no, he is happy with the nine of clubs. And lo and behold, when they open the envelope, inside is a nine of clubs. Also embossed on a plaque in the wallet, in gold letters, is this legend: YOU WILL CHOOSE THE NINE OF CLUBS. Johnny confiscates the wallet and the money.

After the applause dies down, Johnny helps "close all the doors" on this trick by going back over the apparent choices Dan made and the access he had to information about his decisions.

"Now if you were to walk away right now, you might think that that was the cleverest card trick or feat of sleight of hand that you'd ever seen," says Johnny. "But it wasn't a card trick. Were the cards shuffled?"

"Yeah," answers Dan.

"Did you cut them?" asks Johnny.

"Yes."

"Did you deal them faceup and see that every card was different?"

"Yes."

"Did you stop on the card that you wanted to stop on?"

"Pure impulse," says Dan.

"And I saw that you stopped on the only losing card," confirms Johnny. "Did I not offer you a hundred dollars? I begged, implored, even *told* you to go one card further. If you had changed your mind, that wallet would still be in your hands, am I right?"

"Yes," laughs Dan.

SPOILER ALERT! THE FOLLOWING SECTION DESCRIBES MAGIC SECRETS AND THEIR BRAIN MECHANISMS!

As you might suspect, Johnny's retelling of the procedure was actually a clever rewriting of history, one designed to slyly gloss over his suspicious actions. We don't know precisely how Johnny did this trick, because he elected to keep the methods secret. But we can extrapolate from our knowledge of magic to explain how he could have done it.

First, it was Johnny who "shuffled" the cards, not Dan. It is a common sleight of hand to make it appear that a deck of cards is shuffled. So were the cards really mixed up? Probably not.

Second, Dan may have cut the cards, but Johnny made sure Dan cut roughly from the middle. Of course, he omitted this detail from his retelling.

Third, after the cards were cut, Johnny took a furtive glance at the bottom card. This told him the exact order of every card in the deck. How? Because he had stacked it. A *stacked deck* is one in which the magician has carefully placed every card in a predetermined order and then memorized the order. When the deck is cut, the position of only two cards has changed; the rest of the order is preserved.

Fourth, Dan counted out the cards one at a time, starting at the top of the deck, and stopped on whichever card he wanted, right? Not really. Dan was standing in front of a crowd of hundreds of his peers. The possibility that he was going to count out fifty-one cards in the most tedious fashion imaginable was highly unlikely. Instead, Dan counted out seven cards before selecting one—which happened to be the nine of clubs. You can be sure that Johnny knew Dan would not choose the first card, nor would he count out very many cards before selecting. To count out more than about ten would be nerve-racking. Remember, Johnny knew the exact order of the cards in the deck, including the top ten. This means he knew pretty nearly which card was going to be selected, plus or minus five or so cards. Also note that even if Dan had behaved radically and counted out dozens of cards, Johnny could have simply recut the deck, or performed one of many other possible procedures, to force Dan to make the necessary selection in a different way. Since the audience doesn't know the trick, they have no way of knowing if additional procedures are strange or unnecessary. So Johnny was holding all the cards in more ways than one. He could ensure that the card Dan chose was one that Johnny had in his pocket.

Finally, Johnny did not retrieve the wallet containing the matching card and embossed plaque until *after* Dan had made his final selection and presented it publicly. This too was left out of the retelling of the trick.

The fact is that Johnny could have known ahead of time, before he even drove his Cadillac to the event, the narrow range of ten or so cards that were likely to be chosen. He could also force Dan to choose one of the ten cards in a seemingly magical way. If Johnny had ten wallets stored in his suit, all with different cards and embossed messages matching Dan's ten most likely selections, organized so that Johnny could grab the correct wallet in a natural fashion after Dan had made his choice, it would appear as though Johnny had precognition. And in a way he did. He knew exactly how Dan would behave, because Johnny is a master of human observation. Then, by recounting the entire trick slightly inaccurately, leaving out the suspicious bits and distorting certain details, he created false memories for the audience. The creation of such false memories is known in cognitive sciences as the *misinformation effect*—that is, the tendency for misleading information presented after the event to reduce one's memory accuracy for the original event. In Johnny's case, a confidently delivered, coherent-sounding story is much easier to remember than a quick series of subtle movements and visual impressions. In this way, Johnny effectively removed the possibility that the audience, or even Dan himself, could reconstruct the trick and work it out after the fact.

Johnny tells us, "When people see a wonderful piece of magic, they try to figure out how it's done. They have avenues of thought and logic. The magician, just before the denouement or finish, must close all those doors. The only solution is magic."

In 2007, then presidential candidate Hillary Clinton made headlines when she recounted an episode of flying into a United States military base in Bosnia in 1996. "I remember landing under sniper fire," she said. "There was supposed to be some kind of greeting ceremony at the airport, but instead we just ran with our heads down to get into the vehicles to get to our base." Then CBS news aired a video clip from the trip. There was no sniper fire. There was no greeting ceremony. The first lady and daughter Chelsea were seen strolling along, shaking hands, chatting and smiling. Many people had a good laugh at her expense, but Clinton was not lying. Her memory of this particular Bosnia trip had been revised, transformed, and reconsolidated with other memories about Bosnia within the normal circuits of her brain.

Magicians know that memory is fallible and that the more time has elapsed, the worse it is. They count on the fact that your poor memory will not allow you to accurately reconstruct what took place onstage after the fact. Know this about yourself, and keep records of important information and conversations *immediately* after they happen.

False memories can be devastating. Elizabeth Loftus, a psychologist at the University of California, Irvine, and an authority on the malleability of memory, is famous for having shown in the 1990s that some psychiatrists and other mental health professionals implanted so-called repressed (and later recovered) memories in the minds of their patients. For example, one woman, under hypnosis, became convinced that she had memories of being in a satanic cult, of eating babies, of being raped, of having sex with animals, and of being forced to watch the murder of her eight-year-old friend. After later talking with other therapists and realizing that her doctor had manipulated her memory, she sued the psychiatrist for malpractice and won a large monetary settlement.

But for most of us, false memories are prosaic and for the most part harmless. You remember voting in elections you didn't vote in. You remember giving more to charity than you really did. You remember that your children walked and talked earlier than they did. You recall shaking hands with Bugs Bunny (a Warner Bros. character) at Disneyland.

Loftus's studies also explore the misinformation effect. In one example, participants viewed a simulated automobile accident at an intersection with a stop sign. After the viewing, half the people were given the suggestion that the traffic sign was a yield sign. When asked later what kind of traffic sign they remembered seeing at the intersection, those who had been given the suggestion tended to claim that they had seen a yield sign. Those who had not received the phony information were much more accurate in their recollection of the traffic sign.

In another classic experiment, Elizabeth Loftus and her colleague John Palmer asked observers to estimate the speed of a car hitting another, after watching a video recording of a car accident. Observers who were asked how fast the car was going when it *hit* the other car gave lower speed estimates than observers that were asked how fast the car was going when it *smashed into* the other car. Magicians' word choices in recounting the spectators' experiences have a similarly profound impact on their memories of the original events.

Misinformation can change your recollections in predictable and sometimes very powerful ways. You construct a false memory by combining an actual memory with the content of suggestions received from others. During this process, you forget the source of the information. This is a classic example of *source confusion*—something magicians find quite useful.

TYPES OF MEMORY

Your memory feels like a single resource, but this is an illusion. It is composed of subsystems that work in concert to give you the sense of being whole and in command of your past life.

- *Procedural memory*, sometimes known as muscle memory, is for physical skills: skiing, riding a bicycle, shuffling a deck of cards.

- *Declarative memory* deals in facts, and is further divided into semantic memory and episodic memory.
- *Semantic memory* encodes meanings, definitions, and concepts—facts that you know that aren't rooted in time or place: "A horse has four legs," "The capital of England is London."
- *Episodic* or *autobiographical memory* encodes experiences from your unique personal past. This is what allows you to know and recall what has happened to you in your life. The time you discovered someone stole your laptop. The trip to the hospital when your son had an allergic reaction to nuts. Your first magic show.

At a deeper, biological level, all your memories are fallible. The act of remembering an event from your past is not like playing back a mental videotape in your mind's home theater system. It is more like retelling a shaggy dog story that you once heard. You recall a few key phrases and junctures along with the story's overall gist, but you don't recall the exact order of words in the story. When you repeat the "same" story to another person, you reconstruct it in your own way. You freely embellish and fill in missing gaps to make the story flow smoothly. While you might repeat verbatim a few key bits of the original telling, most of the word choices are yours.

Similarly, when your brain lays down a new memory, what it actually encodes is a sparse constellation of personal details and meaningful junctures. When your brain later retrieves the memory, it uses that constellation as a scaffold for reconstructing the original experience. As the memory plays out in your mind, you may have the strong impression that it's a high-fidelity record, but only a few of its contents are truly accurate. The rest of it is a bunch of props, backdrops, casting extras, and stock footage your mind furnishes on the fly in an unconscious process known as *confabulation*.

And it gets stranger. Sometimes a feature that was confabulated during one act of remembering gets reremembered during the next act. In the process, the confabulation can become a permanent feature of the memory. It becomes indistinguishable from the original.

Your memory isn't a partial sketch of the past, it's a sketch of a sketch of a sketch of a sketch of a sketch of a sketch . . . and with every new rendition, more errors can be introduced. Our colleague Joseph LeDoux, a neuroscientist at New York University who studies memory and emotions, says that he used to think a memory was something stored in the brain and accessed when needed. But a researcher in his lab, Karim Nader, convinced him otherwise. Nader demonstrated that each time a memory is used, it has to be re-stored as a new memory in order to be accessed later. The old memory is either gone or inaccessible. Thus your memory about something is only as good as your last memory about it. This is why people who witness crimes testify about what they read in the paper rather than what they witnessed.

Being an expert on the pliability of memory has not prevented Karim Nader from experiencing a memory source confusion shared by millions of other people. Nader, now a laboratory director at McGill University in Montreal, recalls seeing, on September 11, 2001, television footage of the first plane crashing into the north tower of the World Trade Center. But the footage of the first collision aired for the first time the day *after* the attacks. A 2003 study found that a staggering 73 percent of college students tested similarly misremembered the event. "Flashbulb memories"—that is, our seemingly vivid high-definition snapshot memories of traumatic or defining biographical events—are recalled over and over again. Nader's research indicates that the very act of recalling such flashbulb memories can fundamentally alter them.

MEMORY AND THE MEDIA

On February 23, 1981, two hundred armed officers of the Guardia Civil led by Lieutenant Colonel Antonio Tejero burst into the Spanish Congress of Deputies during the process of electing the new prime minister and held the democratically elected government at gunpoint for eighteen hours. The attempted coup d'état ended on the following day, but ask any Spaniard older than thirty-five what he or she was doing at the time of the events and they will be able to tell you right to the smallest detail. The nerve-racking evening and the long night that followed are permanently engraved in their memories.

Or are they? As it turns out, many people remember having *seen* the start of the coup live on TV, as it happened. Not true. Although the coup was broadcast live on the *radio*, the videotaped images were not shown on TV until the next day, long after the coup attempt had collapsed and the hostages were freed.

Spanish writer Javier Cercas writes about this specific memory source confusion in his book *Anatomy of a Moment*: "We all resist the extirpation of our memories, which are the holders of identity, and some prefer what they remember to what happened, so they keep on remembering that they saw the coup live."

Memory source confusion occurs because people are poor at determining the source of information. Remember when the shuttle *Challenger* exploded, killing all astronauts on board, including teacher Christa McAuliffe? Where did you first see the image of those two booster rocket nacelles, flying in lazy figure eights? You remember the image. Who could forget? But did you see it first in the *New York Times*? The *Wall Street Journal*? On the *Today* show? CNN? Was it on TV or did you see it first in the paper? Perhaps it was described to you on the radio? It's difficult to remember, because we don't concern ourselves as much with the source of our information as with the content.

This is why advertising is so effective when it tells us that the product being sold is the best available. Obviously, the source is biased (the company who produces the product paid for the ad). But if we hear it enough, over and over, we eventually begin to believe it. This is one of the reasons that political campaign funding reform is such a hot item: biased advertisements play a huge role in forming our opinions, whether we like it or not, so well-funded candidates have a huge advantage.

At our invitation, Magic Tony is giving a lecture on magic and psychology to some of our fellow researchers at the Barrow Neurological Institute in Phoenix. He has decided that today he will mess with their memories. He will demonstrate how to create a memory illusion by implanting source confusion in an audience.

Tony calls two people, Hector and Esther, to join him in front of the group. He explains that before the lecture started he solicited their help with a trick. He asked Hector to think of a card and to keep thinking of that card throughout the lecture—to simply hold the card in his mind. And he gave Esther a deck of cards and asked her to remove a card and put it in her pocket, without looking at it. So now Hector is thinking of a card, and Esther holds a card but does not know what it is.

The moment of truth. "Hector, what was your card?"

"The jack of spades."

"Esther, look in your pocket. What is the card?"

She pulls out a card: the jack of spades.

Applause. Hector and Esther have astonished looks on their faces. How did Tony do that?

"One of the lovely things about being a magician," he says, "is that you realize words have strong consequences. And this trick is the perfect example of how a magician can use language to create an effect that was not really there in the first place."

Tony asks the audience to think about semantics, ambiguity, and how a sentence can have two different meanings based on context. Consider these two sentences: *I asked him to think of a card. I showed him a card and asked him to keep thinking of it.* They describe the same outcome, except the first sentence implies more freedom, says Tony.

When Tony brought Hector to the front of the group, he used the first sentence to describe what happened. He implanted that lie in everyone's memory. For Hector, who was there for the original event, the misinformation induced a source confusion. He would later remember that he himself chose the card freely. But, in fact, Tony had a different, earlier interaction with Hector. He had fanned a deck of cards and told Hector to stop when he felt the urge to choose one. Once Hector chose a card, Tony told him to keep thinking of it during the lecture. But, as you may suspect, Hector did not freely choose the card. Tony forced the jack of spades. We'll get to forcing techniques in the next chapter. For now, keep in mind that Hector was set up.

Next up, Esther. Consider two sentences: *I handed her a deck of cards and asked her to remove one and put it in her right pocket, and to put the rest of the deck in her left pocket.* Or, *I asked her to pick a card and put it in her pocket.* Again, the first sentence implies a lot more freedom—she has control of the cards—but, says Tony, that is not what happened. He fanned the cards and asked her to select one but, again, her choice was not free. Once again, he forced the jack of spades on her. She was set up, too.

"This is a lame trick," confesses Tony, "but by just using language, it is amplified into a miracle. Hector and Esther each have a false memory simply because of the words I used. By assimilating their memories to match my words, they helped lead the audience straight into a memory illusion and the experience of magic."

CRIMES OF MEMORY

In 1975 an Australian eyewitness expert, Donald Thompson, appeared on a live TV discussion about the unreliability of eyewitness memory. He was later arrested, placed in a lineup, and identified by a victim as the man who had raped her. The police charged Thompson even though the rape had occurred during the time he was on TV. They dismissed his alibi that he was in plain view of a large audience and in the company of other guests on the show, including an assistant commissioner of police. The policeman taking his statement sneered, "Yes, I suppose you've got

Jesus Christ, and the Queen of England, too." Eventually, the investigators discovered that the rapist had attacked the woman as she was watching TV—the very program on which Thompson had appeared. The woman had confused the rapist's face with the face she had seen on TV. Thompson was cleared.

In another illustrious case, the earliest childhood memory of Jean Piaget, the famous child psychologist, was of nearly being kidnapped when he was two years old. He remembered details—being strapped into his pram, watching his nurse defend herself against the kidnapper, scratches on the nurse's face, and a police officer with a short cloak and white baton chasing the kidnapper away. But it never happened. Thirteen years after the alleged kidnapping attempt, Piaget's former nurse confessed that she had made the whole thing up. Piaget eventually realized that his strong visual memories of the episode were fabricated, based on having heard the story told many times by his family.

SPOILER ALERT! THE FOLLOWING SECTION DESCRIBES MAGIC SECRETS AND THEIR BRAIN MECHANISMS!

Magicians plant false memories in many tricks. One notable example is the twisting arm illusion. The magician places his palm on a flat surface and begins twisting it in an impossible 360-degree arc. Then he gives it a second spin. The trick is based on the fact that the magician has twisted his entire arm under his long coat sleeve, which no one can see.* After the first revolution, the magician asks the spectator to do the same. During this effort, he resets his arm for one more revolution. The audience never remembers the magician removing his arm from the table because they are so wrapped up in the illusion. They remember that his palm spun around twice without lifting from the table, a case of misdirection plus false memory.

END OF SPOILER ALERT

*A video of this illusion can be viewed at http://sleightsofmind.com/media/twisting arm.

Memory illusions stem from your need to make sense of the world. If you see a bunch of oranges on the floor and then a picture of a probable cause such as someone reaching for an orange on the bottom of a big pyramid of oranges, you are likely to remember seeing the person grabbing the bottom orange even when you did not. You imagine the event and fill in the details as needed. You can remember events differently from the way they occurred or even remember events that never took place at all.

The crew for the Discovery Channel science show *Daily Planet* on Canadian television had wrapped its visit to our labs, and now the two of us were taking Apollo Robbins to the Phoenix Sky Harbor Airport for his flight back to Vegas. Apollo had come down to our institute for the documentary shoot so we could scan his brain using functional magnetic resonance imaging and also measure the eye movements of people in the audience as he pickpocketed watches and other personal belongings, all for the cameras.* We arrived early for his flight and retired to enjoy some of the best pot stickers in town, at Flo's Shanghai in the shopping concourse of terminal 4 of PHX, and to discuss how magicians not only manipulate the memory of spectators but also use mnemonics to enhance their own memory skills to create magical effects.

Apollo explained that by mentally associating mundane numbers and objects (or people, places, things, activities, concepts, and so forth) with imagined wild caricatures of those things, he could retain the memory of a large number of those associations for an incredibly long time. So long that it didn't seem like memory at all, it seemed magical. He demonstrated by asking Susana to write a list of fifteen items, in random order and hidden from his view, and to call them out as she proceeded. "Number 6: wolf; number 11: market; number 2: roulette . . ." She wrote her list in black pen on one of Flo's white-hibiscus-embossed paper napkins, slightly used, with a Steve's-lower-lip-shaped soy sauce stain on its backside. Here's her complete list.

*The *Daily Planet* video can be viewed at http://sleightsofmind.com/media/Daily Planet.

1 tennis ball
2 roulette
3 bus
4 cookie
5 attic
6 wolf
7 lamp
8 giant
9 fan
10 fingers
11 market
12 hot dog
13 escalator
14 column
15 mirror

Apollo listened but did not appear to be concentrating particularly carefully. At the end of the list, Apollo said, "Okay, now I'll say them back to you in order. Please check them against your list." Susana ensured that Apollo could not see her written list as he proceeded to read them off from his own mental list as promised: "Number one: tennis ball; number two: roulette . . ." He got them all perfectly. He then recited the list backward. Next he asked Susana to cross out seven of the items from the list in random order and to state, out loud, just the number of each entry as it was crossed out. The list remained hidden from Apollo's view as Susana crossed out her selections. Apollo then reported the remaining undeleted items, in numerical order.

Apollo's performance was a straightforward and extremely impressive display of mnemonic power. We reeled under its implications. "How did you do that?" Apollo explained that it was an easy trick that served to boost human memory capacity immensely. "All I did was to associate each number-object pair with an imagined caricature of each object. But the real trick is that I have a list of standard objects that I use to represent each number. It's based on similar-sounding objects, or number homonyms. For example, the number one sounds like 'wand,' so when I make the association between the object and the number, I'm really associating a wand with the chosen object. In this case I burned the image of a tennis ball holding a wand into my memory. Then when the time comes to recite the list, I take each number in turn (backward or forward), recall the associated number-homonym that I always use for that number, and then use that to jog my memory as to the associated object from Susana's list. To delete an item from the list, I imagine each object-number pair being destroyed graphically as Susana crosses it from her list. In the case of the wand-wielding tennis ball, I imagined the pair on fire and then the tennis ball exploding from the internal pressure. I did this for each of the deletions, and then when I went through the entire list in my normal fashion, it was easy to see which numbers had been deleted because I had destroyed them in various ways in my imagination."

As we drove home from the airport, we couldn't believe that we hadn't been trained to do this as neuroscientists. Why weren't we using this technique to give seamless scientific talks, or at least to remember the names of people we met at parties? Why weren't children taught this technique to learn their multiplication tables or other lists of facts? If neurologists could harness these techniques, maybe they could teach Alzheimer's patients to remember better the order in which to don their clothing each morning, maybe enable them to live in their own homes for one more year. It could be a great advance in the treatment of patients with cognitive decline.

We later learned that Apollo used what is called a *peg system*, a form of linking any number of items to a particular digit. Numbers or digits are represented by a word—*wand* for one, *hive* for five, *hen* for ten, and so forth. Then you associate your number word to a vivid visual image. The linking elements are more easily remembered if they interact, are unusual, and tap into your emotions, making you laugh, feel disgust, or perhaps sense danger. Your imagination is what drives the power of the associations. You can also link items without using numbers by associating each word to the next one on your list. For Susana's list, you could think of a giant polka dot tennis ball ricocheting off a roulette wheel and the roulette wheel serving as the steering wheel on the bus, and so on.

These memory systems work because your brain's short-term memory, without some form of assistance, is capable of remembering only seven units (plus or minus two) of anything at a time. After seven items, you begin to forget so as to make room for new items. Or you "chunk" items, as when you remember the seven digits of a phone number (prefix plus four) and an area code. A line of poetry that contains more than around seven beats needs to be broken into two lines.

Another mnenomic strategy called the method of *loci* (the plural of *locus*, which means location, or place), also known as the *memory palace*, has been around for centuries. It's based on the assumption that you can best remember places that you are familiar with, so if you can link something you need to remember with a place that you know very well, the location will serve as a clue that will help you to remember.

According to Cicero, the Roman philosopher, the method was developed around 500 BC by Simonides of Ceos, a Greek poet who was the only survivor of a banquet hall collapse in Thessaly (he had stepped outside). He was able to identify the dead, who were crushed beyond recognition, by remembering their faces based on that day's seating arrangement. Simonides soon realized that he could remember any number of items by setting up walking routes in his mind's eye and visualizing items at various spots along the way. When it came time to remember, he simply retraced the familiar route and easily recalled each item. Unlike linking, loci involves placing a strong visual representation of each item in a geographical space. The nice thing about the method is that, if you forget an item, you can keep "walking in your mind's eye" through the space and pick up the next thing to be remembered.

The method was taken to China in 1583 by an Italian Jesuit priest, Matteo Ricci, who hoped to spread Catholicism but first had to demonstrate the "superiority" of Western culture. Ricci did so by teaching the method of loci to young Confucian scholars who had to learn countless laws and rituals by heart.*

You can try this yourself. Make a list of items you want to memorize, perhaps a shopping list—ice cream, bread, cat food, mayo, chicken breasts, and so on. Now imagine walking through your house or apartment. Start at the front door and make your way through several rooms. (If you live in a one-room apartment, divide the space into distinct areas.) In your imagination, place each item on your shopping list at a single location along your route. Your front door is smeared with Cherry Garcia ice cream. Your living room couch is now a loaf of French bread. The door to your kitchen is shaped like a cat. Your dining room table has dissolved into a mass of mayo. Your bathroom floor is tiled with chicken breasts.

* *The Memory Palace of Matteo Ricci* by Jonathan Spence is a terrific read if you are curious about this historical era.

When you want to remember your shopping list, all you have to do is visualize your front door. You will instantly see the ice cream. As you enter the living room, the French bread will come to mind, and so on. Memory experts say you should make the images as weird and outrageous as possible.

You can also place more than one item in any location. If you have a list of forty grocery items to remember, you could place four items at each of ten locations. Each of these four items should interact at its location. As you open your front door, a gob of Cherry Garcia melts in a loaf of French bread with mayo icing and Meow Mix topping.

THE PERILS OF TOTAL RECALL

Most of us wish we had better memories. But is there such a thing as a memory that is too good? Few individuals (so far as we know) possess near total recall of their autobiographical memories, though no one has yet figured out why. For example, Jill Price, author of a 2008 memoir, *The Woman Who Can't Forget*, says that the days of her life ceaselessly replay themselves in her mind, like a movie running in her head. Give her any date in the past and she can recall what day of the week it was, what the weather was like, what happened to her that day, and the major news events on that date. But she admits that her perfect memory is more of a burden than a gift. She hates change. She cannot forgive herself (or forget) the bad choices she has made in life.

Rick Baron, who also remembers every single thing that ever happened to him, describes his days as "empty." The fifty-year-old has never married and never held a full-time job, but he does compete in occasional trivia contests.

Brad Williams, a news reporter for a family of radio stations in La Crosse, Wisconsin, can also tell you what happened on any date for most of his life. But he, too, talks about the frustrations of having a memory that never lets up.

We, like most married couples, can attest that one of the secrets to a happy long-term relationship is a short memory.

Memory contest champions and many of the world's best magicians use the method of loci. The three-time winner of the World Memory Championship, Andi Bell, can memorize ten randomly shuffled decks of cards in the time it takes him to scroll through them. That's five hundred and twenty cards. Then he can answer any question: What is the thirteenth card in the fourth deck? What is the twenty-second card in the eighth deck? and so on. He never misses a card. Bell's memory palace is a walk around London with specific landmarks. The route and the landmarks—streets, buildings, doorways, traffic lights, mailboxes, and the like—never change. They are fixed in his imagination. Each card has an icon. The jack of clubs is a bear. The nine of diamonds is a saw. The three of clubs is a pineapple, and so on for all fifty-two cards. To memorize a random deck of cards, Bell places an icon at each landmark along the route in his mind's eye. Then he can easily reconstruct the order of the cards by visualizing each icon as he walks through his memory palace.

In an article for *Slate* magazine, the journalist Joshua Foer describes how he entered the USA Memory Championship just to see how he could do. He says competitors insist they are not naturally gifted. They just use mnemonic techniques to help them recall three-hundred-digit binary numbers and to match hundreds of faces with names in twenty minutes.

Magicians and card sharps often use the loci method to stack a deck of cards. A stacked deck, as the name implies, is simply a deck with the cards set up in a predetermined order. It is never shuffled honestly, so that the magician, knowing the position of one card, can always calculate the position of every other card. To memorize a stacked deck, a magician starts with randomly shuffled cards. If you examined them, you would not see anything suspicious. Then the magician creates a personal memory palace to remember the exact order in this particular deck. From then on, he does not shuffle them. He only pretends to mix the cards using a variety of so-called false shuffles. By peeking at the bottom card of the deck as he carries out tricks, the magician can always know the exact order of all the cards by invoking his memory palace.

Stacked decks can also be cyclical and, once you see how they are put together, diabolical. One of the most famous is the Si Stebbins stacking system, originally published around 1898 by William Coffrin, alias Si Stebbins, in a booklet titled *Si Stebbins' Card Tricks and the Way He Performs Them.* To create a Si Stebbins stack, you first organize each playing suit in order. Take all the spades and place them ace, two, three, and so on up to king. Do the same with the diamonds, hearts, and clubs. Then lay these four mini-stacks side by side in the following order: clubs, hearts, spades, diamonds. The mnemonic for this is *CHaSeD.* Now for the stacking. In the stack of clubs, put the ace on top. In the stack of hearts, put the ace, two, three, and four on top. For spades, put the ace through the seven on top. For diamonds, put the ace through the ten on top. Then make a full stack by piling the little stacks of clubs, hearts, spades, and diamonds. You now have a stacked deck. You can cut the cards any number of times, and by looking at the bottom card you can always know the card on top. How? By counting. The stacking results in the fact that every card is three values higher than the preceding card.

HOT READING

Some corporate magicians use amazing memory feats to appear to read minds. For example, when they receive a list of people attending a seminar at a given company, they can Google the names to find a subset with photos posted online. Then they memorize the face and name of each one along with any personal information they can gather. (Before Google, such magicians could look up people in newspaper archives at the library, or even send accomplices to discover information at the company office.) The amount of data collected can be quite large. During the corporate seminar, the illusionist can then claim mentalist powers and "read the minds" of various people by providing their names, work and home addresses, office and home phone numbers, children's names, pets' names, genealogical information, etc. The goal is to provide so much detailed information that it seems impossible that the magician could know it all in advance, and that the only solution must be that he's reading the mind of the client in real time. In the world of magic, such subterfuge is called a *hot reading*. But the real feat is that the magician did indeed remember all that information, and was able to conjure it as if by magic during the seminar.

We saw another type of stacked deck in action at the Magic Olympics in Beijing. Juan Tamariz, the famous Spanish magician, called a volunteer from the audience and, after much joking, prompted him to "pick a card, any card." The deck looked normal, but it really contained only six cards—the three of hearts, the nine of clubs, the seven of clubs, the jack of diamonds, the two of spades, and the ace of hearts—repeated in order over and over. Tamariz fanned the cards in front of the volunteer and noted the exact position of the card that was selected. By counting down the line of cards, Tamariz was able to identify and then surreptitiously lift an identical card from the stacked deck. While the magician did not know the identity of the chosen card, he now had an exact copy of it in his possession and was able to produce it, as if by magic, at the end of the trick.

END OF SPOILER ALERT

A little later, Tamariz demonstrated a two-part trick involving memory that stunned the assembled experts. By now you probably realize that it is extremely difficult to fool a magician. They know every sleight of hand in the book and are constantly on the lookout for misdirection, fake shuffles, clever props, and the like. One false move and they're on to you. Tamariz began his trick with an incredibly corny routine. He announced he would teach us some comedy. Pacing the stage and wringing his hands, he asked everyone in the audience to touch their two index fingers together, making a horizontal line in front of their eyes, and then stare into the distance. "You see?" he said. "You've created a magic sausage floating in front of your eyes. And if you're really hungry, you can use six fingers to make three sausages."

The magicians in the enormous lecture hall were stumped. What was Tamariz talking about? Sausages? Fingers? Just then, Tamariz jumped into the audience and corrected the finger position of a guy in the front row. "You're doing it all wrong!" he screamed. Then he praised the man in the next seat over. "That's perfect! It's so good you can slice them up and share." With that, Tamariz did a karate chop in the air, through the perfect finger sausages, and produced a string of three large kielbasa.

Um, what was that all about? The magicians squirmed with concern. Poor old Tamariz, he must be losing his touch. Of course, the sexagenarian began prancing again and pulled off several gorgeous tricks flawlessly. Everyone forgot the sausage nonsense.

About forty-five minutes later, Tamariz invited a woman onto the stage and had her count out ten cards. He had her place a rubber band around them, carry them to a table across the stage, and return to his side. Next he invited a man to step up onto the stage next to the banded cards. The two volunteers were a good fifteen feet apart, and Tamariz never left the woman's side. Tamariz asked her to count out another ten cards onto a table and then hold them in her two hands. With much fanfare he proclaimed he would make some of the cards teleport across the stage. Once he was done, there should be thirteen cards in a stack on the other side of the stage. Tamariz magically waved his hands toward the woman and asked her to give him the cards so he could count them for all to see. Only nine cards remained. One was missing. He handed her the cards again and repeated the magical wave. Now he recounted, and two cards were missing. A third time . . . three cards were missing. "Let's see how I'm doing," he said, and asked the man to count out the cards next to him. The man did and said, softly, "Um, there are ten cards here."

Tamariz pretended to be crushed. "Ten? You only have ten? Are you sure? Could you count them again?" Yes, only ten and not thirteen. Tamariz was deep in thought. "Ummm, could you check your left pocket?" Nothing was there. "Your, umm, right pocket?" Nada. People started shifting in their seats. Everybody wanted to disappear.

"Could you check your inside left breast jacket pocket?" Tamariz said. Still nothing.

Dejectedly, Tamariz said, "And your inner right jacket pocket?" The man's left hand entered his right inner jacket pocket and he looked up suddenly in surprise. He stopped cold. Hackles rose on one thousand necks. Slowly, the man removed his hand from the pocket. In it were three cards.

"Three cards!" screamed Tamariz. "Three cards! It's a miracle!"

Only you know it isn't. He planted the cards on this guy during the sausage trick, now a long-forgotten ruse. (The onstage volunteer was the same guy that Tamariz corrected right before pulling sausages out of the face of the man on the next seat.) And no one—neither the volunteer nor the world's best and brightest magicians—remembered that he had had that opportunity an hour earlier. Memory can play tricks on us all.

8

EXPECTATION AND ASSUMPTION

How Magicians Make ASSes of U and ME

It's not easy to follow magic acts by the Great Tomsoni, Apollo, Teller, and James the Amaz!ng Randi. But Mac King, the last speaker at our 2007 Magic of Consciousness symposium in Las Vegas, is undaunted. Connoisseurs will tell you that Mac is one of the most influential magicians in the world. He is also one of the nicest and without a doubt funniest.

For his stage act—which he presents twice a day, five days a week, at Harrah's casino on the Las Vegas strip—Mac affects the persona of a country bumpkin. The first words out of his mouth are "Howdy! I'm Mac King." The audience shouts back, "Howdy!" A bit of a Danny Kaye look-alike, Mac wears outrageously tasteless plaid suits that somehow accentuate his beak nose and reddish blond hair fashioned in a classic bowl cut. He guffaws and giggles as he performs his routines. Mac is having such a great time you can't help but be drawn in.

Mac explains the source of his mirth. Each day, he calls new people up on stage to assist with a trick, and each time he finds something funny and spontaneous to say or do with them. In this way Mac never lets his act grow stale, which is undoubtedly why his show is one of the highest rated in Las Vegas.

Mac is also an inventor of new illusions. He often creates tricks for other magicians' shows and is always on the lookout for inspiration. His office is littered with props. You get a feeling for his devious sense of humor and inventiveness through a story he tells about a trick he pulled on his wife. Several years ago, Mac purchased two pencils with little hands on their ends instead of the usual erasers. The hands were about two inches long and rubbery with wires that could be used to oppose the fingers. "I thought they were funny," says Mac. One day, looking at the pencils on his desk, Mac had an inspiration. He called to his wife, who was in the bathroom. "Honey, could you please fill the sink with hot water? I want to shave." She obliged and then stepped into the shower. Mac came in after her in his bathrobe. He splashed the water, hid his hands in the sleeves of his robe, and stuck out the tiny rubber hands. Then Mac let out a bloodcurdling scream. When his wife looked out from the shower, she saw Mac standing there with the tiny shrunken hands over a hot steamy sink. "She freaked out," says Mac with a satisfied expression. This is not a trick he can take to the stage in Vegas but it illustrates the way his mind works.

At our science conference, Mac performs one of his favorite tricks from his show at Harrah's. But first he explains a cardinal rule in magic: never perform the same trick twice, at least for the same audience. "It's really hard," explains Mac, "because if you do a trick that really fools people, they'll say after it's over, 'That was the greatest thing. Do it again! Do it again!' You think to yourself, okay, why not? What's the harm in doing the same trick over?" He gives a conspiratorial smile. "Well, I'll tell you. If you see a trick a second time, there are lots of clues."

To demonstrate, Mac picks a guy from the audience to come onstage. His name is Marvin Chun. He's a famous visual science professor from Yale, but Mac doesn't know that. And Marvin certainly isn't the one teaching the lessons today. "Marvin," says Mac, "there's a surprise! You get a prize for helping. I keep the prize in my shoe. Let's see what I have for you today, Marvin." Mac removes his right shoe and a packet of honey falls to the floor. It's not a very appealing prize, and so everyone titters. "No, really, Marvin, you get my shoe," says Mac. "These are Rockports. Really nice shoes. You know how you tell a real Rockport shoe?" Mac tips his shoe over. "They have these big rocks in them."

An impossibly giant rock falls to the floor with a loud thud. Unlike the honey packet, the rock is a huge surprise. You have no idea where it came from and how it got into his shoe. "I had it in a secret hiding place," Mac says in answer to the unspoken question.

MISTAKES

From our angle onstage, behind the performers, we don't have a good view of the rock, but we learned of its existence during Teller's talk when Mac accidentally dropped the rock from his back pocket. It made a loud thud that everyone in the room must have heard, but only those of us at the dais realized what happened. We'll never forget the mirth and chagrin on Mac's face as he retrieved his rock unceremoniously, on all fours, and looked up at us as he groped under his chair. Even though no one in the audience saw the rock, you would think the loud noise, followed by Mac's undignified crawling around the stage, would have been a clue to whatever he was up to. But the sound seemed to fall on deaf ears. As far as we could tell, no one heard it, and no one seemed to remember seeing Mac crawl under his chair.

It occurred to us that magicians, like all of us in our jobs, must make mistakes all the time. But since a magician's mistakes involve unlikely objects and actions, most spectators do not realize their significance. Magicians know this, and it gives them the courage to simply keep going even in the face of glaring logical errors. Indeed, one of the hallmarks of a good magician is the ability to recover smoothly and seamlessly from unexpected mishaps. Mac told us a hilarious example of this from early in his career. One of his signature tricks involves pulling a live goldfish out of his mouth and dropping it into a glass of water held by a volunteer from the audience. Don't worry, he says, the fish is not in there very long—only a few seconds—or it could not survive the heat and saliva in his mouth. Anyway, the first time Mac did this trick onstage with a volunteer, he started choking as "that little fish decided to swim down my throat. I tried to hack it back up. Then I turned around and threw up bits of my sandwich from lunch and the fish into the suitcase I keep on stage. The guy next to me said 'Eeeeewww!' but no one else reacted. I keep an extra fish in case of a fish accident and so I recovered and finished the trick." Mac's eyes widen. "Later, no one asked me 'Did you vomit onstage?' Everyone saw it. It's so weird. I don't know what's going on in people's minds."

The lesson here is to keep moving forward after everyday mistakes. Even though magicians make mistakes all the time, they put them behind them, keep moving forward, and the audience hardly ever notices. You should do the same. Just like a magician, continue to glide along as if nothing happened, and your mistake will go unnoticed most of the time, too. Don't get mad. Don't get embarrassed. Just reset yourself as best you can and go tuck a fresh goldfish from your suitcase into your mouth.

SPOILER ALERT! THE FOLLOWING SECTION DESCRIBES MAGIC SECRETS AND THEIR BRAIN MECHANISMS!

The rock weighs about five pounds and is the size of a papaya. To demonstrate that doing the same trick twice is a mistake, Mac performs it again, exactly the same. This time you can see more clearly how he does it. Mac tips his shoe and the honey packet falls out. Classic misdirection. But you are on to him. Instead of focusing your attention on the honey packet, you see him reach into his back pocket and slip the rock into his shoe. When the rock falls out with the loud thud, you are no longer surprised.

Mac asks how many people saw him slip the rock into his shoe, and about half the hands in the room go up. "I'm so happy that some of you noticed," Mac says. "I've been a little worried that it wouldn't get easier for any of you the second time!" And then Mac asks if he should do it again, a third time. Murmurs of assent. But this time

Mac changes the method and produces another surprise. He does not reach for his back pocket to bring out the now familiar rock. Instead, he simply tips the shoe and shakes it, and when nothing comes out he reaches in and—pulls out a huge rock! Only this time it turns out to be a sponge rock. He had it stuffed in his shoe the whole time.

END OF SPOILER ALERT

Mac's demonstration illustrates how apparent, but not actual, repetition is a powerful ally to the magician. You become habituated to seemingly repeated actions and gloss over the details. For a magician, the devil is in those details. The audience has a deep-seated bias to assume that effects that look the same are done in the same fashion. It's human nature.

In your everyday life you deduce how things work by observing them repeatedly. Hammers drive nails. Cups hold liquid. Microwave ovens heat food. You don't have to think about them. Magicians use this habit of your mind against you to hide the method behind many of their tricks. They know that when you see an effect repeated—the rock drops out of the shoe—you naturally assume that the repetition is accomplished by the same method. But then comes a surprise when the rock is made of sponge.

Mac used a different method on the third iteration to throw the audience off track, eliciting a big surprise. Remember the story from chapter 2 in which Danny Hillis fooled Richard Feynman day after day with the same trick, not because Feynman couldn't correctly guess at the method after a day of pondering it (he could), but because Hillis kept changing the method, and so Feynman's explanations were demonstrably wrong. Feynman was flummoxed by apparent repetition.

Using apparent repetition, a magician can deliberately raise suspicion about a possible method, and then at the very end show you that the only theory you've got is wrong. This principle, known as the Theory of False Solutions, was formulated by Juan Tamariz, the Spanish magician in the crazy hats introduced in chapter 5. Johnny Thompson calls it "closing all the doors," which means reducing all of the possible explanations of an effect down to none, until only impossible (magical) explanations remain.

SPOILER ALERT! THE FOLLOWING SECTION DESCRIBES MAGIC SECRETS AND THEIR BRAIN MECHANISMS!

The whole point of apparent repetition is to set up false expectations. The magician shows a trick using method number one, and you form a theory of how he did it. Next, he apparently (not really) does the same trick again, but wait—now that you are watching for the telltale sign that your theory is correct, you can see that your theory is impossible. Hmm. Okay, you form a new theory. The magician does the trick again (no, he doesn't really, it just looks as if he did), oh, drat, your second theory is wrong, too, because now that you try to see if your second theory holds water, you can see that the magician is not hiding the card on the back of his hand (though that's exactly what he was doing the second time).

The magician is one step ahead of you, setting up expectations with each iteration and then crushing them just as you begin to understand.

END OF SPOILER ALERT

One of the greatest weapons magicians have going for them is that your mind operates via prediction. To grasp the meaning of this, imagine what you "knew" as a newborn infant. You could root for a nipple and stick out your tongue, but otherwise the world was mostly a background curtain of meaningless sights, sounds, and sensations. You could feel the pull of gravity and sense light and dark patterns, but nothing made sense. You were not even aware that you had a body. It's doubtful you could be called conscious the day you were born.

Fortunately, infants and babies rapidly race out of this twilight to build up representations of the outside world, their bodies, other people, their feelings and emotions. Every experience is carved into the developing brain's neural circuitry via *plasticity*, the lifelong ability of the brain to reorganize itself based on new experiences. In this way, each person builds up models of what to expect throughout life.

Early on, you learned that the feet and hands you liked to put in your mouth were your own; you taught yourself to roll over, sit up, crawl, and walk until your movements became engrained in areas of your brain that plan and carry out movements. Then you walked, ran, and—if you practiced a lot—played a sport without thinking about or planning the required motions. But now imagine you are walking down a city street and are so engrossed looking up at some signs that you don't notice a six-inch curb is just ahead. Your distracted brain predicts that the sidewalk is flat while the road is six inches lower. You take a step with the exact precision needed for your foot to land on the sidewalk. And what happens? Surprise! Your foot crashes into the roadway. You failed to predict a common feature of an ordinary walkway.

Early in life you learned to recognize faces and voices. You figured out how to manipulate adult caretakers to get what you needed. If you had nurturing parents, you learned that your cries would be met with love and attention. If you had emotionally unstable parents, you learned that your cries might be met with indifference or punishment. If you had parents who experienced good days and bad days (who hasn't?), you learned how to cope with emotional ups and downs. Most importantly, you learned what to expect from intimate relationships in your life long before you could talk.

You learned to speak based on expectation. Toddlers extract the meaning of their native language from a stream of syllabic sounds and gradually become proficient in vocabulary and syntax. Thus if someone says to you "how now brown," your brain will predict the word "cow" in a flash. But if the person instead says "wolf," your prediction fails and you are surprised.

The same principle applies to vision, hearing, touch, and all your cognition, including your beliefs, which are, after all, constructs of your learned predictions. In other words, perception is not a process of passive absorption but of active construction. When you see, hear, or feel something, the incoming information is always fragmentary and ambiguous. As it percolates up the cortical hierarchy, each area, having its own specialized set of functions, analyzes the data stream and asks: Is this what I expect based on my very last experience? Do I need to fill in some of the gaps in the data stream? Does it jibe with my other past experiences? Does this conform to what I already know about the world? Your brain is constantly comparing incoming information to what it already knows, expects, or believes. Every experience is measured up against prior beliefs and a priori assumptions.

Indeed, all great art is based on violation of prediction. When you go to the movies, you see the same twenty plots unfold over and over. Often the film is boring because it's predictable. But a talented director challenges your predictions. You are surprised, entertained. The same goes for painting, poetry, novels, and great magic acts.

Alas, the automaticity of predictions can get you into hot water. For example, NASA put commercial airplane pilots into a flight simulator and asked them to do a set of routine landings. On some of the approaches a huge commercial aircraft was plopped on the runway. One-quarter of the pilots landed on top of the airplane. They never saw it because they had been led to believe that there was nothing unusual and the runway would be clear.

James the Amaz!ng Randi is a short man with a long Santa Claus beard and a gigantic personality. He's the guy who showed that Uri Geller's spoon bending could be done with mundane methods and who introduced us to Teller and other leading magicians. Randi commands the room wherever he goes. It is no wonder that he plays the role of elder statesman for the American magic community. As founder of the James Randi Educational Foundation, he protects society at large from charlatans and frauds of the paranormal. The foundation offers a one-million-dollar challenge to anybody who can prove paranormal powers of any kind. After more than twenty years and numerous challenges, no one has qualified to collect the money.

Randi moves slowly up to the podium at our Magic of Consciousness symposium. He's getting on in age, but the twinkle in his eyes is youthful and beguiling. Randi explains that you will easily accept unspoken assumptions and that you tend to believe information that you learn for yourself as opposed to being told it. Prediction at work.

"My purpose here today is to show you that audiences will easily accept their own assumptions, but not assertions made by the conjurer," says Randi. "In other words, when we tell them something is so, they have good reason to doubt us because we're there to trick them. So we should try to allow them, as much as we can, to assume things. Conjurers do well to take advantage of the spectator's misplaced confidence in his own ability to arrive at a correct solution."

Randi demonstrates. "I have already deceived you folks," he says. "When I walked onstage, you assumed I was talking into this mike." He pushes away the large microphone affixed to the lectern. The real mike is tiny, clipped to the top of his lapel. "Why did you believe it? If you were asked specifically 'Did he use the house amplification system,' you'd say yes, he did. And you wouldn't be telling a lie when you reconstructed the experience for others later on. You'd be telling what you believe to be true. But it wouldn't be true."

Another example: "Many of you think I'm looking straight at you. But no, I'm looking at a blur of faces out there. I can't see you, because I normally wear glasses with corrective lenses." Randi removes the glasses from his head and pokes his fingers through empty frames. "Now why would someone come up before you wearing empty frames? What's the use of that? To make a point, ladies and gentlemen." The point being that people don't question lies that have no reason to be lies.

But why don't people question unspoken assumptions? The reason is that such assumptions have already been questioned and established as fact. As children, we pulled our grandparents' glasses off of their faces, stuck them in our mouths, and tested the lenses with our tongues. As adults, we feel no further need to continue to lick the glass. We've become habituated to the fact that glasses frames have actual lenses in them. But this is just an observation, not an explanation. It is critical to go further into the neuroscience here and ask how the brain actually accomplishes habituation, and why.

The *why* is easy: thinking is expensive. It requires brain activity, which takes energy, and energy is a limited resource. More important, thinking takes time and attention away from other tasks, like finding food and mates and avoiding cliffs and saber-toothed tigers. The more you can safely file away as established fact, the more you can concentrate on your current goals and interests. The less you wonder whether somebody's glasses frames actually contain glass, the better off you are.

Habituation is created through a neuronal process called *synaptic plasticity*. Eric Kandel of Columbia University recently won the Nobel Prize for his work establishing this process in a little-appreciated sea slug called an aplysia. Kandel recorded from a variety of neurons in the aplysia's nervous system while blowing air onto the slug's gill. Aplysia don't like air puffs on their gill, so they retract it. But air puffs aren't really harmful, and retracting the gill is tiresome and burns precious calories, so as the air puffs are repeated, the aplysia habituates and eventually stops retracting the gill in response. The neural signals concerning the air puffs become more and more minute until neurons eventually stop signaling the air puff altogether. That's synaptic plasticity, and it's the neural mechanism of habituation. We humans do the exact same thing as the lowly sea slug, only we do it with more fancily processed perceptions and behavioral options. We don't question whether every pair of glasses we see contains glass, because experience has taught us that we can safely assume they do, and the synaptic pathways responsible are habituated to that fact. There's no longer a need to lick the glass.

Once you've habituated to a feature of the world, it becomes a humdrum and seemingly immutable part of the fabric of life. Stable, reliable, unchanging. That's why magicians prefer to rely on unspoken assumptions over explanations whenever possible.

Mentalism, the subject of the next chapter, is a specific branch of what magicians call the conjuring art, and its success relies on the audience's assumptions. "Mentalism deals with things that are apparently extrasensory, precognition, divination of various kinds, but it's all a form of conjuring," says Randi. "There is nothing to it in reality. They are tricks. You see, the mentalist does very well by allowing his audience to assume things."

Mac King is standing in Susana's lab at the Barrow Neurological Institute during a documentary shoot for the Australian Broadcast Company's weekly *Catalyst* science show. Two cameras are rolling: one was brought by the show's producers and one is ours. Max

Maven told us that Mac was the very best there is at tossing coins from one hand to the other, so here he is. It's not that he drops them less than most people. It's that Mac can toss a coin through the air only to have it disappear upon landing. You see it clear as day. Mac tosses the coin up in the air one, two, three times in his right hand, then tosses it to his left hand. You see it fly through the air. His hand closes to catch the coin, and then opens wide to show that the coin is gone. Incredible.

Here's how it works. Mac is in fact tossing the coin vertically in his right hand. But when he makes the toss to his left hand, his right thumb surreptitiously holds the coin in his palm and stops it from flying. So he's only pretending to throw the coin. The left hand closes as if the coin is in flight and "catches" it. But of course the coin was never there—so why do you see a coin flying through the air?

The trick takes advantage of an implied or inferred motion illusion stemming from the motion-sensing portions of your brain. First, a visual region of your brain that tracks the movement of objects or targets in space and time—called the *lateral intraparietal*, or LIP—receives information about the actual motion of Mac's right hand. Neurons in this area predict the trajectory of the flying coin based on his hand movements. Then, when Mac's right hand abruptly stops, the motion-selective neurons of two other visual areas (the primary visual cortex and a specialized motion-sensitive visual area called MT) sense the motion of Mac's left hand closing. A major component of this trick is that Mac closes his left hand at the same moment that the coin would have flown through the air had it actually been launched. Without his closing his left hand, there is no motion for the motion areas of your brain to detect. Without the closing of the left hand, the trick is much more likely to fail. But because motion information of the closing hand closely matches the implied motion of the fake-launched coin from the right hand, the predicted trajectory of the illusory coin jibes with the actual trajectory of the left hand's closing fingers, and your brain is satisfied—incorrectly!—that the coin actually flew into the left hand. In fact, only Mac's fingers were moving.

Similarly, have you ever pretended to throw a stick for your dog during a game of fetch? The dog spins around and starts to take off, expecting the stick to follow the implied trajectory. This is because circuits in her brain that are active during the perception of real motion also respond to implied motion. Her brain tells her there is a stick in flight and off she goes.

END OF SPOILER ALERT

During a magic act, you are as easily duped by implied motion as your dog. Take the vanishing ball illusion, a dumbfoundingly simple trick. A magician throws a little red ball into the air three times. On the first two throws, he catches it in his hand. But on the third throw the ball mysteriously vanishes. You "see" it go up and then it disappears. Amazing.

The magician pulls this off by capturing your gaze with joint attention. Each time he throws the ball, he moves his head up and down to exaggerate the ball's trajectory. But on the third toss, he only pretends to throw the ball. He hides it in his hand while his head moves up to track the ball's apparent path. But you, a slave to social cues, move your head up along with his. And that is when you get the sudden sensation that the ball—which you think you've been following with your eyes—has disappeared in midair.

Gustav Kuhn, a psychologist and magician at the University of Durham in the UK, showed a film clip of the vanishing ball illusion to thirty-eight students and carefully tracked their eye movements as they watched. Two out of three reported that they saw the ball leave the magician's hand on the third toss.

The eye tracking revealed that the students were not looking at the point in space where they thought the ball vanished. Rather, the magician used his gaze to covertly direct their attentional spotlights to the predicted position of the ball. His eye movements overruled what the students' own eyes were seeing. The illusion works in part because your brain pathways for eye movement and perception operate independently, and in part because you have low-resolution vision outside the center of your gaze, meaning that you are not surprised that you don't see the ball as it is thrown. Your attention follows the presumed trajectory of the ball because of the magician's gaze. Once you catch up to the ball with your eyes, it literally disappears, because now you can see with your high-resolution central vision that the ball is gone. It proves again that the direction of your gaze can be separated from attention.

The effect may be related to the same kind of *representational momentum* we saw in Mac King's tossed coin trick—the final position of a moving object that suddenly disappears is perceived farther along the path of motion than its actual final position. If so, the neural correlate of the effect lies in area LIP of your posterior parietal cortex.

The vanishing ball also illustrates *priming.* You are more likely to see it vanish in midflight after real tosses have primed you to know what an actual tossed ball looks like.

Priming is a powerful force in everyday life, by which subtle suggestions made to your subconscious mind can influence your subsequent behavior. Try this: Answer the following questions out loud and quickly. Don't stop to think about your answer. What color is snow? What color are clouds? What color is whipped cream? What color are polar bears? What do cows drink? If you said cows drink milk, you were primed by the previous questions to choose something white (cows drink water, Farmer John).

Psychologists are fond of studying priming in laboratory settings. Here are a few examples of some recent experiments.

- Subjects were asked to read a list of words related to old age and infirmity—*wrinkled, gray, nursing home, dementia*—interspersed with neutral words. Afterward, they walked more slowly toward the campus elevator than did others who did not read such words. The effect did not last long, but the change in their behavior was noticeable.

- Chinese female students took a math test after filling in ethnic or gender information. Being reminded of their gender resulted in lower test scores (the gender stereotype is that girls are bad at math), whereas being reminded of their race resulted in high scores (Asians are good at math).
- Half of the participants in another study were subliminally primed with the words "Lipton Ice"—repeatedly flashed on a computer screen for 24 milliseconds—while the other half was primed with a control that did not consist of a brand. Priming the brand name Lipton Ice made those who were thirsty want the Lipton Ice. Those who were not thirsty, however, were not influenced by the subliminal message, since their goal was not to quench their thirst.
- Advertisers use priming to trigger consumption of junk food. In an experiment, elementary school children watched a cartoon that contained either food advertising or advertising for other products. While watching, they were given goldfish crackers. Kids who saw the food advertising ate 45 percent more crackers.

PRIME MENTALISM TRICKS

Magicians, especially mentalists, often use priming to bias your responses. For example, here is a mentalist trick normally done with either three or seven spectators, but it should work on you just as well. Please get a pen or pencil and follow these instructions, in the order they are presented, and follow them as quickly as possible.

1. Choose a number between 1 and 50.
2. But there are a few rules to your choice.

3. The number must be double-digit.
4. Both digits must be odd numbers.
5. One number must be larger than the other.

Write it down quickly.

Okay, now we are reading your mind. Look at the number and concentrate hard on its value. We're starting to pick up your thoughts. Once we have the number solidly, we'll write it into this book and send that manuscript off to the printers so that you can read it here.

You guessed the number 37. Yes? Yayy! We read your mind backward in time. No, we got it wrong? Well, clearly, either you didn't follow the instructions or you didn't concentrate hard enough. Maybe you should go buy another copy of this book and see if it works better with that one.

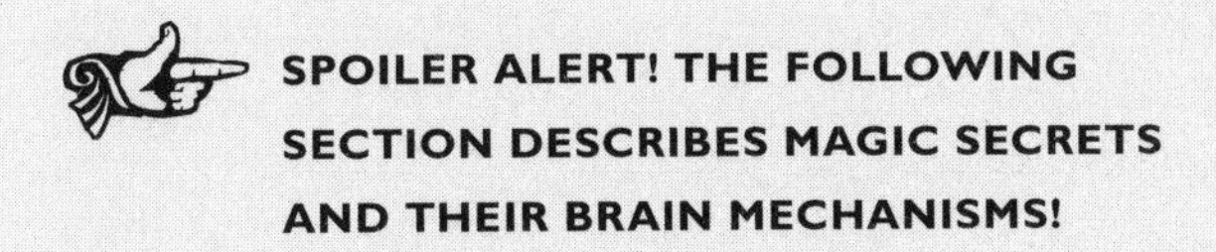

SPOILER ALERT! THE FOLLOWING SECTION DESCRIBES MAGIC SECRETS AND THEIR BRAIN MECHANISMS!

Want to know how the trick works? In the first place, nobody really knows. But here's what we do know. We reduced the number of choices by requiring a double-digit number. That narrows it down to between 10 and 50. Then we said both numbers must be odd. That leaves only ten choices, between 11 and 39. Then we said you couldn't have duplicates, leaving you with just eight choices: 13, 15, 17, 19, 31, 35, 37, and 39. Okay, narrowing fifty choices down to eight leaves us with a 12.5 percent chance of getting the guess correct, which is better than our original 2 percent chance, but still quite low. So why do people tend to choose 37? Well, we primed you to think about 3 and 7 by starting off our discussion by saying that the trick works best on groups of three or seven. That's not true. The trick is generally done on one person, not on a group. There are certainly other contributors to why this trick works, since it still works most of the time even without the priming, but the other factors are currently not well understood.

END OF SPOILER ALERT

Priming can also lead to perceptual misinterpretations in the form of expectations gone wrong, which can get you into serious trouble. For example, our colleague Peter Tse at Dartmouth College served as an expert witness in a recent case of a man who shot at what he thought was a bear and killed a man instead. According to Tse, the twenty-one-year-old hunter was primed to see a bear. He had seen his younger brother kill a bear earlier that day and he wanted to bag one for himself. The victim was out picking berries in the Vermont woods during bear hunting season without the reflective orange clothing that hunters wear to avoid shooting each other. (This is almost grounds for the Darwin Award, but no matter.) The hunter saw the bushes shake, took aim, and, seeing a bear in his sights, sent a bullet through the victim's shoulder, both lungs and his heart, and out the other shoulder. The berry picker was dead in less than a minute.

The hunter and his brother fled the scene once they realized their mistake. Their uncle later talked them into turning themselves in. The verdict was negligent manslaughter with a one-year prison term.

Tse made his case on the idea that priming—seeing the brother's successful kill earlier that day—had lowered the hunter's ability to detect a false alarm, which in this case meant the erroneous detection of a bear. It was, he said, an example of *signal detection theory.* Signal detection theory was invented during World War II to help determine when British radar operators should scramble fighters to shoot down German bombers. False alarms were bad because if you scrambled fighters to defend against a nonexistent attack, the country was left vulnerable to a real attack from a different direction for a lengthy period of time while fighters flew home, landed, reserviced their planes, rested the flight crews, and prepared for another scramble event. False alarms were expensive and dangerous. On the other hand, failing to scramble at the earliest possible moment might mean that bombs would fall in the heart of London. Scientists refer to this kind of mistake as a "miss." The question was how to determine ideal criteria for minimizing both false alarms and misses. And how do the radar operators set their own internal criterion for deciding when a blip on the screen is a Nazi bomber?

In the case of the bear hunter, he was single-mindedly determined to kill a bear that day. Never mind that nobody holding a gun should be single-minded about anything—the fact is that our desires lead us to see what we want to see. The hunter's ability to detect a bear was heightened to the maximum level, but this same criterion also heightened his ability to mistake a man for a bear. He was in the perfect trigger-happy mood to act on a false alarm. In the end, that's exactly what he did, and it all came down to how he handled the inevitable tension between false alarms and misses.*

Like priming, your tendency to hold biases and stereotypes makes false alarms more likely. For example, Keith Payne, a psychologist from the University of North Carolina, Chapel Hill, asked people to sort guns of various kinds from hair dryers and caulk guns and other gun-shaped tools. He used a bias measurement technique championed by Harvard psychologist Mahzarin Banaji. With this method, the level of a person's bias (racial, gender, or other) is determined by measuring their reaction time to concepts that conflict with their belief system. Payne found that American experimental subjects linked black people more easily to guns, whereas they associated white people with tools.

This stereotype turned lethal in 1999 when a twenty-three-year-old African student, Amadou Diallo, was killed in New York City because he reached for his wallet when police ordered him to halt. In his country of Guinea, you are supposed to take out your wallet when approached by police. Diallo was shot at forty-one times and hit nineteen times. The cops claimed they saw a gun, not a wallet, and were acquitted, resulting in riots.

Given that false alarms are prevalent, what can we do to decrease their occurrence? One idea is to manipulate the observer's expectations. This is the philosophy of the Transport of London campaign to get drivers to be more aware of cyclists on the road. Car drivers are constantly on the lookout for other cars, but they often miss bicycles and motorcycles. Transport of London uses gorilla-in-our-midst-like demos in television commercials in an attempt to increase driver awareness and reduce the likelihood that drivers will hit a cyclist. It

*A Google search using the terms "shot" and "mistaken for" produced a host of gunshot victims (over 3 million results), including those mistaken for coyotes, turkeys, monkeys, deer, foxes, and, in the case of one unfortunate snorkeler, a giant rodent.

should work. In the Simons and Chabris gorilla demonstration, people are more likely to see the gorilla if you tell them that there could be a gorilla in the movie.

SPOILER ALERT! THE FOLLOWING SECTION DESCRIBES MAGIC SECRETS AND THEIR BRAIN MECHANISMS!

Magicians use both bias and priming to cause false alarms, which relates back to Tamariz's Theory of False Solutions. Recall that one way to create strong misdirection is to give clues that a certain method is being used to accomplish a trick when in fact it's a different method altogether. Well, magicians also use prior biases to accomplish false detections. Remember Mac King's fake coin toss and Kuhn's disappearing ball? When you see the coin and ball tossed in the air for real, it serves to plant the bias that the magician always tosses the object. In these tricks, the magicians use repetition to increase your bias toward making a false alarm (detecting a coin or ball when none is there), but also to decrease the possibility of your missing an actual coin toss. Imagine a card sharp playing three-card monte—an ancient confidence game in which the victim bets he can find a target card among three facedown playing cards. The magician gives the observer several trials to see where the target—say, the queen of spades—correctly lies. This increases the victim's confidence and shifts the criterion (the victim's sensitivity to the position of the queen) up. Then *wham!* the card sharp uses sleight of hand to swap out the queen, causing a miss during a trial with a large bet.

END OF SPOILER ALERT

When our son Iago was two, Steve showed him a magic trick. Steve felt he had gotten pretty good at the trick and wanted to show off. But Iago was unimpressed. Here was a kid who was endlessly delighted and entertained by the fact that he could blow out a candle but found something utterly impossible to be utterly banal. You already know why. His brain was still naive enough about the laws of physics and causality that he had no predictions on which to base a sense of surprise. He was still young enough that we could show him how to make an object travel through a magic space-time wormhole and he would simply note it, and maybe play around with this new fact for a while—in exactly the same way he played around with pouring liquid from one container to another, or pulling his socks on and off, on and off—and that would be that.

Mac King agrees with us. Kids are harder to fool, he says, because they don't have strong expectations about the world. They just think magic exists. Some people really can make a coin dematerialize. If you believe in Santa Claus, what's not to believe in a magic show? It's just a bunch of adults magically transporting a coin around or making cards disappear into thin air. What they really want to see is something difficult and funny, like a triple somersault resulting in the seat of the jumper's pants splitting up the middle.

Randi agrees. Children are notoriously difficult to deceive, he says, because they're not sophisticated enough to be fooled. They have not built up bulletproof models of probability and impossibility.

Thus we can ask: When does a child's mind reach a level of maturity that allows her to be delighted or amazed by a magic trick? How does she acquire expectations? Indeed, what do babies know? When do they learn to predict the world? When are their expectations violable?

Such questions raise a deeper quandary. When infants are born, how much of their brains are preloaded for acquiring knowledge about the world? Are their brains blank slates, or do they possess innate structures that are locked and ready to absorb knowledge? In the 1920s, the Swiss developmental psychologist Jean Piaget pioneered this inquiry and concluded that infants younger than nine months have no innate knowledge of the world. He said they have no sense of *object permanence*—the idea that a thing can exist even when you don't see it. Piaget also argued that babies gradually construct knowledge from experience, including the capacity for empathy, which he suggested came rather late in development.

Modern cognitive neuroscientists challenge many of Piaget's conclusions and assume that infants are born with some knowledge of the physical world. They are "statistical learning machines" who have a rudimentary capacity for math and language. Young babies have everyday ideas about psychology, biology, and physics.

Because babies can't talk, developmental psychologists have devised numerous strategies for gleaning information about infant cognition. In "baby labs," infants sit in high chairs or on their parents' laps and observe simple scenarios. The experimenter then measures how long an infant looks at one object compared to another or at a series of events. The idea is that their gaze reveals how interested they are in the object or if they can detect something out of the ordinary—indications that they have simple models of how the world works. For example, babies may become less interested when they see the same event happen over and over. They grow bored. When a new event comes along, they will look longer at it, as long as they notice the difference.

Elizabeth Spelke, a developmental psychologist at Harvard University, has carried out scores of experiments on the reasoning abilities of children. In one, Spelke showed that babies as young as three and a half months will look longer at impossible events (such as a hinged wooden panel moving through a box) than at possible ones. They have, she says, a basic understanding of physical events that appear to violate gravity, solidity, and contiguity.

Such research also shows that infants have a sense of object permanence far earlier than Piaget postulated. In an experiment, babies watched a toy car move down an inclined track, disappear behind a screen, and reemerge from behind the screen farther down the track. Then the researchers put a toy mouse behind the track, raised the screen, and rolled the train again. No problem. Finally, they put the mouse on the track, lowered the screen, then secretly removed the mouse and rolled the train. Infants as young as three and half months looked longer at the possible mouse-crushing event, suggesting they had a sense of object permanence. They knew the mouse existed, and they knew it was located where the train should hit it.

David Rakison, a psychologist at Carnegie Mellon University in Pittsburgh, also uses toys to explore what babies know. Rakison studies infants' abilities to categorize objects. You might think young children naturally lump cows and horses in one group and cars and planes in another. But does that mean they know what the objects are? When Rakison removed legs and wheels from such toys, the babies put cows and cars together. He notes that infants can tell that dogs are different from cats when they are three months old, but they do not know that dogs and cats are alive until they are three years old.

Our son Iago first saw a giant tortoise at the age of eighteen months, during a visit to the Phoenix Zoo. The enormous animal (the size of a kiddie pool) was stationary for a long time, and then it started to walk laboriously toward us, along the fence. Iago exclaimed "Vroom vroom," as if encouraging the reptile to move faster. Lacking any experience of tortoises, he had simply decided that the strange approaching object was some sort of slow-moving car.

Other researchers use animals or dolls to explore what is called the *theory of mind*—the innate ability of one person to sense the state of mind of another person. A great example of theory of mind in chimps was presented in the 1999 Scientific American Frontier television program *Animal Einsteins*. In this episode, Alan Alda was dressed up like a veterinarian, wearing scrubs and a mask, as he marched into the chimpanzee enclosure at Georgia State University. He held what looked like a spear, a one-meter-long metallic post with a huge needle on the end. His host, Sue Savage-Rumbaugh, knew that this getup would definitely get the attention of her chimps. It was the same outfit any one of the veterinary staff would wear when they entered the enclosure with the intent to give a shot to one of her animals. She had a "theory of mind" concerning her chimps: that they would see a syringe-wielding vet and be very unhappy about it. She was right.

As Alan walked down the caged hallway, a chimp sitting above the chain-link ceiling, several animal holding cages farther down the hall, watched him like a hawk. A second chimp was released from its holding cell behind the enclosure and entered the chain-link cage. This second chimp could see the first chimp above the hallway outside the cage, but because of a cage separation wall made of brick, the second chimp could not see Alan. But after seeing a signal made by the first chimp, the second chimp stopped in his tracks and looked at the separation wall as if it were the devil himself. He clearly knew that something evil his way came. And this was possible only because the first chimp knew that the second chimp could not know that Alan was coming, and so he signaled him. This incredibly complex behavior shows that the first chimp had a theory of mind. He knew that the second chimp had a mind and that he could not possibly know about the impending doom. So he warned him.

The famous Sally-Ann test is used to look for the emergence of theory of mind in young children. A child is introduced to two dolls, Sally and Ann, and is shown that each doll has her own box, with a candy or toy hidden inside. Then the child is told that Sally is going out for a minute. The experimenter removes the Sally doll from the scene, leaving her box behind.

Next, the child is told that Ann is going to play a trick on Sally. Ann opens Sally's box, removes the candy, and hides it in her own box. Sally returns, unaware of what has happened. The child is asked where Sally will look for her candy. A child with a theory of mind will realize that Sally doesn't know that Ann has played a trick on her. She predicts that Sally will look in her own box for the candy and discover it is missing. But a child lacking a theory of mind will see the situation based on what she knows in her own mind to be true and will predict that Sally will look for the candy where it actually is: in Ann's box.

Very small children tend not to guess correctly in this test, since theory of mind takes time to develop. Most children get it right by age six or seven, although some three-year-olds are capable of it (our son Iago, three years and seven months old at the time of this writing, failed the test).*

Babies and young children also differ from adults in the styles of their attention, their ability to lie, and their sense of time. In her book *The Philosophical Baby,* Alison Gopnik, a developmental psychologist at the University of California, Berkeley, explains that in order to focus attention, you need strong input from your *prefrontal cortex*, which is the last brain area to develop in humans. With the help of mature circuitry, your attention works like a narrow spotlight, focusing on one thing at a time. In babies and young children, Gopnik says,

*Theory of mind is critical to magic because magicians know you have it and that it is a lever by which to control your mind. It's the basis of joint attention. Magicians overtly attend to objects and locations of potential interest in order to control your attention, drawing it away from a secret move.

attention operates more like a lantern, casting a diffuse light on its surroundings.

"We sometimes say that adults are better at paying attention than children," writes Gopnik. "But really we mean just the opposite. Adults are better at not paying attention. They're better at screening out everything else and restricting their consciousness to a single focus."

"Adults can follow directions and focus, and that's great," says John Colombo, a psychologist at the University of Kansas. "But children, it turns out, are much better at picking up on all the extraneous stuff that's going on. And this makes sense. If you don't know how the world works, then how do you know what to focus on? You should try to take everything in."

These ideas are consistent with the neural correlates of attention that we discovered in collaboration with Jose-Manuel Alonso's lab, described in chapter 4. Attention results from the activation of inhibitory neurons, which in turn suppress neurons in the surrounding visual regions that could cause distractions. When and where you focus your attention, you are also suppressing the potential surrounding distracters. The harder you concentrate, the larger your central attentional activation and surround suppression become. Gopnik and Colombo's studies suggest that babies and children don't suppress surrounding distracters as well as adults do.

In an experiment by John Hagen, a developmental psychologist at the University of Michigan, children are given a deck of cards and shown two cards at a time. They are instructed to remember the card on the right and to ignore the card on the left. Older children and adults direct their attention to the card on the right and remember it. But young children often remember the cards on the left, which they were supposed to ignore.

Gopnik also argues that children under five experience a different sense of time. The world is less ordered. They forget what happened a minute ago and how they felt. They don't seem to anticipate their future states. They don't project what they will think and feel later on. They don't have the concept of logical, internally driven thought.

But kids above the age of five have started to develop a sense of consecutive time and a stream of consciousness that flows in an unbroken stream with a unified self at the center. Magicians need these functions in order to make magic magical. Without these processes, there is no strong sense of cause and effect, and therefore no inviolable rules that can be violated. Before you're five years old, your entire life is a magic show, so what's one more trick? We have often asked magicians how early a child can understand and enjoy magic. The usual answer is five years old.

So what kinds of magic tricks do appeal to children younger than five years of age? What would surprise our little Iago? We decided to ask one of the world's premier children's magicians, Silly Billy, aka the New York City performer David Kaye, how he deals with this issue. Not surprisingly, he says the magic tricks that work with children tap into a child's basic knowledge about the world.

For example, pulling a coin from a child's ear is deeply magical. Kids have had ears for their entire lives. They use them to listen and learn. But producing money is not a familiar characteristic of the human ear. So, says Kaye, when a magician pulls a coin from a child's ear, it is magic.

The needle through the balloon is another trick that works on kids. They know balloons can pop. They know a needle will pop a balloon. So when the magician inserts a needle into a balloon and it does not pop, the child sees it as magic.

If you take a crayon and rub it on a surface, it leaves a mark, says Kaye. But the Magic Drawing Board trick, developed by Steve Axtell, goes way beyond the familiar. The magician draws a face on a large board. Suddenly the eyes start moving, the mouth opens and closes. The face becomes animated and carries on a conversation with the magician. This breaks natural laws related to drawing with a crayon.

If you hold a cup of water and turn it upside down, water will spill out. But in the Slush Powder trick, says Kaye, a magician pours water into a Styrofoam cup and turns the cup upside down and the water has vanished. Similarly, a magician can make a cone out of a newspaper, pour milk into it, unfurl the paper, and show that the milk is gone. Kids go wild when they see this.

Finally, if you put small items into a container and move the container across the room, the objects will still be inside. Young children know this. But when the magician places an object in a "change bag" and says the magic words, the object is no longer there. This is magic that a child can believe in. None of it requires a theory of mind.

That a magical feeling takes place when our expectations are violated is a fact that makes perfect sense to anybody who considers it. What is surprising is the horsepower necessary for the brain to form an expectation in the first place. It takes years of constant study for children to develop proper expectations of the world around them or even of the people they have loved their entire lives. This is despite the fact that, like adults, children possess the most powerful known computational device in the universe—the human brain—to guide them in developing expectations. The fact that magicians can collapse those long-held expectations like so many houses of cards is incredibly useful in developing new neuroscientific methods to find out the exact architecture of not only our adult minds but also the developing minds of our children.

9

MAY THE FORCE BE WITH YOU

The Illusion of Choice

James the Amaz!ng Randi is back onstage, only this time he's at the Naples Philharmonic Center for the Arts in Florida. He's doing us a favor by performing several mentalism tricks at the 2009 Best Illusion of the Year contest.*

Mentalists are magicians who use mathematical probabilities, human nature, sleight of hand, gimmicks, and trust to make it appear that they can read your mind. Their acts are highly theatrical, often invoking "mystical" powers of clairvoyance, telekinesis, telepathy, precognition, divination, and mind control.

Unlike many New Age psychics, who claim to possess supernatural powers,† mentalists such as Randi, Max Maven, Derren Brown, and other top performers do not lay claim to paranormal faculties.

*See http://illusionoftheyear.com.

†The magic community is divided between magicians who claim to have psychic powers (they will explicitly claim supernatural abilities as part of their performance, or implicitly lead the spectators to that conclusion) and those magicians who make no such claims. Unsurprisingly, the two magical traditions don't see eye to eye. The magicians featured in this book are strong supporters of the no-claim tradition, and openly admit to performing "tricks" in their acts.

Rather, their illusions are spun from an ability to exploit human gullibility and, as you will see, to carry out brilliantly sneaky, underhanded maneuvers.

Today Randi is performing a book test. In this act, the magician may ask a volunteer to exercise free will in picking out a magazine, finding a random word somewhere in the magazine, and thinking about the word silently. The magician divines the word by reading the volunteer's mind.

Randi looks out into the audience, hand shielding his eyes from the spotlights like a sailor blocking the sun as he peers out to the horizon. "I met a young woman outside before the show who agreed to assist me with this next trick. Could you please stand up?" A young woman stands near the center of audience. Randi introduces her as Zoe.

"Now, before we get started, could you please confirm that we have never met before tonight?"

"Correct," she says.

"That you are in no way being coerced by me, that you haven't been paid by me, and that any decision you may make has not been given to you by me?"

"No," says Zoe.

"When we met in front of the hall tonight you chose a word from a magazine completely at random and of your own free will?"

"Yes."

"Was that magazine a different copy of this specific magazine, which we chose from the rack of free literature outside this very building?" Randi says, as he pulls a folded free apartment rental guide from the breast pocket of his navy blazer and slowly opens each page to show the audience that there is lots of text.

"Yes."

"And I asked you, did I not, to open that magazine to any page you wanted having lots of text on it, and to choose any word you liked from that page freely, while I stood with my back to you?"

"Yes."

"And then you destroyed the magazine, correct?"

"Yes."

"It would be impossible for me to know what word you chose, right?"

"Right."

"Okay. You have a piece of paper with that word written on it. Could you please circle that word now, as I try to read your mind?"

"Okay," she says, and she circles the word on the page.

Then Randi begins to pace. He prowls to stage right and to stage left. The shadow he casts from the spotlight jumps in animation against the pleats of the red velvet curtain standing a full two stories high behind him. His brow knots severely as he rubs his forehead and temples. He mumbles to himself in a slightly disconcerting but amusing fashion.

Finally, Randi stops in front of an easel holding a large writing pad next to the podium. He uncaps a huge black Sharpie and, with his eyes closed, looking up into the lights, right hand pressing on his eyelids, left arm extended with unsheathed pen ready to strike, he speaks. "I'm starting to get something," he says as he writes an *N* on the paper. "It's all coming now." He proceeds to receive mental vibrations for eight more characters as well, spelling out the phrase: **NθI+d3)3P**.

Finished, and visibly exhausted from the effort, Randi pulls his hand from his face. He looks at the pad for a long time, totally silent, then turns back to the crowd. The throng starts to fidget as they become embarrassed for the poor old coot.

"Is the magazine written in the English language?" Randi eventually asks, failing to hide the disappointment in his voice.

"Yes," giggles Zoe, as other nervous laughs arise from the audience. Zoe is still standing, and she is so embarrassed for Randi that when she responds she has to lower the paper she has been using as a mask to hide her face.

"Are you a mathematician?" Randi hopes sadly.

"No," says Zoe.

"Okay, well, I guess I didn't get it," Randi concludes, shoulders and chin slumping. "What was the word?"

"Deception," says Zoe.

"What? Hmm? I'm sorry, I didn't hear you," says the suddenly frail octogenarian, bent over to bring his now cupped ear closer, eyes squinting into the glare.

"Deception!" yells Zoe.

"Hmm. Yes, well . . . sometimes these things fail," he says dejectedly. Looking up at the pad one last time, he does a double take and says, excitedly, "Oh, but wait a minute! I think I see what happened!" Now thirty years younger, he positively leaps as he lifts the page from the pad and rips it off. He turns to the crowd with the ripped page and slowly rotates it 180 degrees as he says, "I must have gotten the signal from you upside down and backwards!"

Once the rotation is complete, the page reveals the now legible message: **dε(εP+IθИ**.

The crowd roars as Randi receives his standing ovation.

The next morning, Randi returned to his home in Fort Lauderdale, the James Randi Educational Foundation, or JREF. Susana and I were thrilled to drive him on the two-hour jaunt back from Naples. Take our word for it, we've traveled all over the world with magicians, and in the summer of 2009 we even flew, drove, and sailed across China with two hundred Spanish magicians, so we know: if you ever feel like taking a boisterous road trip, go with a magician.

The JREF serves as a skeptical third party, rooting out fraud and outrageous claims made by psychics, faith healers, hypnotists, and even deluded scientists. We arrived at the foundation building, a refurbished house surrounded by peacocks, in time to find the staff celebrating the news that they had just sold out the next The Amaz!ng Meeting (TAM), to be held in London that fall. We were shown to the Isaac Asimov library, the foundation's extensive collection of magic literature, with books that line every side of a large windowless wood-trimmed conference room complete with a huge central conference table that would be the envy of any CEO. Notes and paraphernalia pertaining to Randi's next book, *A Magician in the Laboratory*, were strewn across the desk.

On the road trip, Randi had told us that the magazine test he did on Zoe is one variant among many for a classic trick known as the Book Test. "Every mentalist does one," Randi says in the library. "It's fundamentally an illusion of choice."

"Allow me to demonstrate. My dear," Randi says to Susana, "if you would please pick any book you like from the shelves." Susana comes back with a randomly chosen magic book and shows it to Randi. "Good, good," he says, "but let's make sure it doesn't have too many pictures. You need to have plenty of choices of text." He takes the book and flips rapidly through the pages. "Okay, great," he says, as he hands back the book. "That book will work nicely."

"Now I'll choose a book of approximately the same size," he says, grabbing another tome from the shelves. "Next I'll read your mind, but first you need to choose a page somewhere in the middle as I flip through the pages of this book." He holds the back of the book flat in his left hand as he lifts the cover and all the pages of the book to a forty-five-degree angle. He flips the pages down with his thumb in a cascade, and about halfway through Susana says, "There."

"Page 174," says Randi. "Now, let's review. You chose a book of your own free will, correct?"

"Yes."

"You chose the page you wanted, right?"

"Correct."

"Now you will freely choose the word you want from that page in the book you are holding," Randi says.

"Uh-huh," Susana confirms.

"So there is no way for me to know what word you are about to choose, right?"

"Well, I'm sure you will, but I don't see how you will do it!"

Randi chuckles, "Well, that's where you're right! Now, my dear, please open your book to page 174 and pick a word from the top line. Don't choose an article or some unsubstantial word, pick a nice, meaningful, beefy word."

Susana flips to page 174 of her book, reads the first line, picks a word—"stellar"—and you already know what follows.

But how does he do it? Randi can't know what word Susana is about to choose, can he? Randi explains that book tests are an illusion of choice because the choices are known to or forced by the magician. In this case, Randi's retelling of the trick and Susana's choices are—well, we don't want to say *dishonest*, but they are not quite accurate. Let's go over them. First, Susana does indeed choose her own book. No forcing there. But does she choose page 174? Not really. Randi is the one who flips the pages, not Susana, and Susana never actually sees which page is showing when she says "Stop." Randi is lying when he tells her it is page 174. So now the question is, how can Randi know the first line of page 174 for a book Susana randomly chose? Has he memorized the first line in every book from the thousands in his library? No. When Randi flipped rapidly through the pages of Susana's book to "check for pictures," he wasn't really looking for pictures. He was looking for a glimpse of any page in which he could resolve both a word from the top line and also the page number from the upper corner. It just so happens he saw "stellar" on page 174 as the pages flew by. It's challenging with all the blurred movement because he flips the pages quite fast. But with practice it can be done, and Randi needed only the single word and its page number to make the trick work.

How does Randi know exactly which word Susana will choose? He doesn't. But there are only so many big beefy or stellar-like words on a single line of any normal book. Even if Susana happens to choose a different big word, Randi can recover by saying, "Oh, but the word 'stellar' is in fact there, isn't it? You must have unconsciously found the word 'stellar' to be more interesting than the other word you chose, and that's why I picked it up more in your brain waves." Randi uses mentalism tricks to restrict her choices to a single word or just a few possible words. So when he "reads her mind," he is actually just making an educated guess that has a low probability of failure. And in the event of a failure, it's an easy one to fix.

END OF SPOILER ALERT

Let's go back to the illusion contest on the previous night. Randi's just received his standing ovation for reading Zoe's mind. But how did he do it? What's the method behind this particular book test? Zoe's choices seemed essentially infinite. Has the Amaz!ng Randi (he's told the members of the audience that since they are all friends now they should call him by his first name, Amaz!ng) somehow actually divined the word "deception" from all possible words, and in spectacular fashion to boot? No, Zoe is definitely being fooled. She may feel that she has thousands of secret choices and is being directed by nothing other than her own free will, but that is not the case.

Again, Randi's retelling of history is, well, telling. It's true that Zoe chose a word from a magazine that was found outside the philharmonic hall. That's Randi's version of "found art." He likes to use local literature because it makes the illusion seem all that much more convincing, since he could not have prepared anything. And in a way, he didn't. He relied on his wits.

It is also true that Zoe scanned the magazine without Randi's seeing her do it, and she circled the word ("so as not to forget it," Randi had told her) with *his pen* before ripping out the page and discarding the magazine in a trash can. But wait—Randi announced during the show that the magazine had been "destroyed," not discarded. An important modification, to be sure, but not enough of a misstatement that Zoe would complain. For most people, when an object enters a trash can, it ceases to exist and is for all intents and purposes destroyed. But no act is beneath the Amaz!ng Randi. Once Zoe entered the philharmonic hall, Randi did indeed go dumpster diving to recover that ripped magazine. Zoe had torn the relevant page from it, true, but now he knew which page was missing. And because Zoe used Randi's own specially selected pen to circle the word, a nice hard ballpoint pen, the circling of the word left an impression that was barely discernible on the adjacent page of the magazine. So Randi knew the page and its exact location. To find the word itself, Randi took a second copy of the magazine, ripped out Zoe's chosen page, put it under the embossed page from Zoe's magazine, and lined up their corners so that they overlapped perfectly. Randi then poked a hole through the embossed circle, marking the word Zoe chose on the page below. It was, of course, "deception."

In yet another incredibly devious move, Randi took a third pristine copy of the magazine and ripped out the same page Zoe had ripped out, with the same tear pattern, before putting it in the trash can to replace Zoe's original. This new copy had never been touched by the pen and so it had no embossed circle in it. If Zoe, or some other person in the audience, reconstructed Randi's methods and came back to do a little dumpster diving of their own, they would find what looked like Zoe's original ripped magazine. They would remain mystified.

To further throw off the audience, Randi had Zoe circle the word during the act itself so that if any of the other audience members saw the ripped page during or after the act, they would assume that the circle had been created during the show and not before. (Randi was careful not to mention that the word had been circled before the show.) Randi similarly implied that Zoe had written down the word on a piece of paper rather than saying that she had the page ripped from the magazine, so that people wouldn't even think of trying to get the evidence and reconstruct the trick.

Randi allowed Zoe to make her word choice in truly free manner, but it was not a secret choice, though it felt like one to everybody, including Zoe. Randi had controlled her every move from the minute he said hello. Then, all he needed to do was figure out how to spell "deception" upside down and backwards. For a master magician like Randi, that little bit was the hardest part of the whole trick.

Since watching Randi perform, we have investigated other mentalist tricks to see what they reveal about human nature. Here are three of our favorites.

In the 1089 Force, the magician first asks you to pick a three-digit number whose first and last digits differ by two or more. Let's say you pick 478. You write it down. Step two, the magician asks you to reverse the order of the number and write it down: 874. Third, you are asked to subtract the smaller number from the larger number, in this case $874-478=396$. Fourth, reverse that number to become 693 and add it to 396. Your answer is 1,089.

So far, so good. Now the magician hands you three or four books (or more if he wants to lug them around). You choose one, any one, your free choice. The books look normal, not marked in any way. He says, "Excellent choice! Now turn to page 108 and look at the first line. Count over to the ninth word and hold it in your mind. Got it?" You follow his instructions. The word is "yellow."

"Concentrate now," says the magician. "I am going to read your mind. The word is coming into focus, slowly, slowly. I see a, hmm, a color? It starts with, let me see, it starts with a *y*? Yellow! The word is 'yellow.' Am I right?"

Indeed he is. The ninth word at the top of page 108 in the book you picked is "yellow." He memorized it before the show. He also memorized the ninth word at the top of page 108 in all the other books. If you had chosen any one of them, he would have known the word you'd find.

The 1089 Force is a mathematical trick based on the fact that any three-digit number manipulated in this manner always—*always!*—adds up to 1,089. The magician simply picks the books and looks up the word he wants you to find. He could, for example, tell you to turn to page 10, count down to the eighth line, and look up the ninth word in that line (1089.) The effect is astounding and always entertaining.

Another mathematical force convinces you that everyone in the room can be made to share the same mental picture. You all are asked to think of a small number and then silently perform the following operations. Double the number. Add 8 to the result. Divide the result by 2. Subtract the original number. Now convert this number into a letter of the alphabet (1=A, 2=B, 3=C, 4=D, and so on). Next, think of the name of a country that starts with this letter. Got it? Now think of an animal whose name starts with the next letter. Finally, think of the color of that animal.

The magician makes a dramatic pause. "Oh, my, your collective image must be wrong. There must be a problem. There are no gray elephants in Denmark." The trick works because everyone must choose a country that starts with *D*, and Denmark is the most common. The next letter is *e*, and most people think of an elephant. And who isn't going to think of a gray elephant?

People usually make the same choices because when they are

asked to stand up and speak in front of hundreds of other people, they tend to say the first thing that comes to mind. Mentalists know that the number of countries starting with *D* is vanishingly small, and that the likelihood that they'll pick the Dominican Republic is low unless they are either unusually cool under fire or have some time to consider. Most people then choose "elephant" and not "emu" for the same reasons. They are nervous. They're scared of looking stupid in front of so many people, and they can't think clearly enough to come up with something clever.

Mentalists may also use something they call the "one-ahead principle": to give the impression of reading your mind, they stay one step ahead of you at all times. The coincidences are multiplied in your mind, resulting in the illusory feeling that the only explanation is supernatural ability.

Magic Tony showed us a trick based on this principle. He gave us a deck of cards to shuffle thoroughly and then he spread the deck facedown on a table and announced that he would predict our choices. "First, you will choose the nine of hearts," he said. We slid a card out of the spread, Tony looked at it, and set it aside.

Without showing it to us, he exclaimed, "Good job! Now I predict you'll choose the two of clubs."

We chose another card at random and slid it to him, still facedown. He looked at it and said, "Excellent!"

Tony gathered the remaining cards and shuffled them. "Now you will choose the queen of spades. "Pick any card as I run my thumb down the corner of the deck by saying 'stop.' " He held the deck in one hand and riffled his thumb down the deck.

About halfway through the deck we said, "Stop."

Tony removed the card, picked up the other two cards we had chosen, and turned over all three in front of us: the nine of hearts, the two of clubs, and the queen of spades. Wow!

To accomplish this trick, Tony first surreptitiously memorized the card at the bottom of the deck: the nine of hearts. He then spread the cards facedown and asked us to make our choices, announcing that we would choose the nine of hearts.

When we picked the first card, we thought it must be the nine of hearts (after all, this was a trick by a terrific magician) but we could not verify that with our own eyes. In fact, the card was the two of clubs, which Tony saw with his own eyes.

Then Tony announced that for our next card we would choose the two of clubs. (Hmmm, he already had that card on the table but we followed his direction and pulled another card. He saw that it was the queen of spades.)

Tony then collected the remaining cards, did a false shuffle so as to keep the nine of hearts exactly where he wanted it, and asked us to choose a card as he riffled the deck with his thumb. We chose a "random" card in the middle of the deck, but he lifted the cards from where he was keeping the nine of hearts while distracting us from the sleight with eye contact. He removed the nine and laid it out with the other two chosen cards to show that his three predictions were correct. In fact, he had simply "predicted" whatever card had previously been chosen.

END OF SPOILER ALERT

You get the idea. Mind reading is a setup, flimflam, bunkum, even treachery—but why does it work again and again? Why are you so taken in? Why do you entertain a nanosecond of belief that a magician could even begin to have this ability? How does he force you to follow his will?

Forcing is a method used by magicians to make you think you've made a free choice when in fact the magician knows in advance exactly what you will do—what card you'll choose from a deck, what word you'll choose from a book, what object you'll choose from an array of items on a table. He is in complete control. When a mentalist has you in his clutches, your sense of free will is an illusion.

A classic method of forcing is called *magician's choice.* You are asked to make a free choice among items but, no matter what you choose, the magician calls the shots by how he verbally responds to your choices.

SPOILER ALERT! THE FOLLOWING SECTION DESCRIBES MAGIC SECRETS AND THEIR BRAIN MECHANISMS!

For example, if the magician puts two cards facedown on the table and wants you to choose the one on the right, he will say "Choose either one." If you choose the one on the right, he goes on with the trick. If you choose the one on the left, he will say, "Good, you keep that card and I'll use the remaining one." Thus he forces the card he wants.

END OF SPOILER ALERT

A "force" is not unlike the cinematic version you may have seen in George Lucas's original Star Wars movie. There's a scene in which the Jedi master Obi-Wan Kenobi and our hero Luke Skywalker, along with robot sidekicks R2-D2 and C-3PO, are trying to leave the planet Tatooine. En route to the spaceport they are stopped by two armor-clad, gun-toting imperial storm troopers. Obi-Wan gives a sly smile and wave of his hand as he tells them, "These aren't the droids you're looking for." The storm troopers appear confused. One of them parrots back, "These aren't the droids we're looking for." Obi-Wan dominates their minds, forcing them to believe and say whatever he tells them. After the storm troopers wave our heroes past the checkpoint, Obi-Wan explains to the young Luke Skywalker, "The Force can have a strong influence on the weak-minded."

Except that in the real universe we are all weak-minded, and magicians are the Jedi masters.

Forcing works because your brain is on a constant, active lookout for order, pattern, and explanation and has a built-in abhorrence of the random, the patternless, the nonnarrable. In the absence of explicability, you impose it. When you think you are choosing something, but the choice is changed on you or distorted in some way, you nevertheless stick to your guns and justify your "choice." You confabulate.

Confabulating is a fancy term for shamelessly making things up. It is another of those potent and ubiquitous brain processes that occur all the time but to which you are seldom wise. Normally this process is beneficial. For instance, confabulation is what allows you to "see" people and objects in drawings instead of the tangle of dark lines that you are actually looking at. It is also what allows you to "see" faces in clouds; it allows your perception to be flexible and creative. But when this sort of pattern imposition goes on at higher levels of cognition, the implications can get a little uncomfortable. Your mind will go to surprising lengths to preserve its sense of agency and choice and continuity of the self. When you are influenced by others, you rationalize their influence as being good decision making on your part.

The breadth and depth of confabulation is revealed following some kinds of brain injury, when the mind's normal system of checks and balances is perturbed. For example, when the right brain hemisphere is damaged, spectacular delusions can arise about the state of the body. Here is Dr. Anna Berti, a neuroscientist at the University of Turin in Italy, interviewing one her patients, "Carla," whose paralyzed left arm rests in her lap next to her good right arm.

"Can you raise your right arm?"

"Yes." Carla's arm goes up.

"Can you raise you left arm?"

"Yes."

The arm remains motionless. Berti tries again.

"Are you raising your left arm?"

"Yes," says Carla. But still her arm does not move.

"Can you clap your hands?"

Carla moves her right hand to the midline of her body and waves it in a clapping motion. The left hand is motionless.

"Are you sure you're clapping?"

"Yes."

"But I can't hear a sound."

Carla replies, "I never make noise when I do something."

Insistent denial of paralysis was long thought to be a psychological problem, Berti says. It was a reaction to a stroke: *I am paralyzed, it is so horrible, I will deny it.*

But it is not a Freudian dilemma. Rather, it is a form of so-called *neglect syndrome* in which a brain area involved in the mental simulation of movements, the *supplementary motor area,* is damaged. When you close your eyes and simply imagine a golf swing or skiing motion, you activate this part of your brain.

When Berti asks Carla to raise her left arm or clap her hands, the region that imagines such movements produces a familiar pattern of activity in her brain. But the regions that carry out those movements and also maintain awareness of making them are not working.

The conflict is overwhelming. Carla's sense of having moved via simulation is powerful. Awareness is absent. Paralysis is complete. Her brain's solution: confabulate.

If prodded, patients make up stories to explain their lack of action, Berti says. One woman said her arm "went for a walk." A man claimed that his motionless arm did not belong to him. When it was placed in his right visual field, he insisted it was not his.

"Whose arm is it?" Berti asked.

"Yours."

"Are you sure? Look here, I only have two hands."

The patient replied, "What can I say? You have three wrists. You should have three hands."

Neuroscientists can also unmask your confabulatory nature in the laboratory. Two young Swedish scientists have developed a new scientific method that uses magic techniques to examine the fascinating way in which confabulation operates in the intact, healthy, ostensibly rational brain.

We are in Benasque, Spain, nestled in the heart of the Pyrenees, at the Pedro Pascual Center for Science, a retreat designed to bring together scientists from every discipline to hash out ideas in hopes of inspiring new interdisciplinary approaches. Miguel Angel, the Spanish magician whom you met in chapter 5, has just completed his demonstration of change blindness. Now up on stage are two neuropsychologists from Sweden, Petter Johansson and Lars Hall, from Lund University. These two twentysomethings are today's fair-haired boys of cognitive science, and not just because they're Swedish. They have brought a veritable smorgasbord of methods to the discipline. One especially sweet meatball was featured in an October 7, 2005, article in *Science* magazine describing the invention of a new and powerful method for studying human cognition, rationalization, and decision making called *choice blindness.* And they did it using magic.

Johansson explains that their experiments were inspired by the so-called *introspection illusion.* Introspection, he says, does not provide a direct pipeline to your unconscious mental processes. Instead, it is a process whereby you use the contents of your conscious mind to construct a personal narrative that may or may not correspond to your unconscious state. When you are asked to say why you have a particular preference or how you arrived at that preference, your personal self-report of your internal mental processes is confabulated. To put it bluntly, you are unaware of your unawareness.

Johansson and Hall describe their incredible experiments in a fast-beat tag-team style. They show a short movie of themselves, made by the BBC the previous year, to illustrate their new approach. It starts with one of them displaying two photographs of two young women to either male or female subjects. The images have been previously matched for attractiveness, so the women are more or less equally good-looking. When they hold up the photos, the subject, seated across the table, points to the one he or she deems more attractive. Next the photos are placed facedown on the table and the selected photo is pushed across the tabletop to the subject, ostensibly so that he or she can pick it up and examine it more closely. "Here, take a closer look and tell us why you chose it!" the researchers ask, entreating each subject to consider the reasons leading to their choice. Johansson and Hall run the experiment dozens of time on each subject and dutifully record the considered opinion of each beauty judge, each time with a new attractiveness-matched pair of photographs.

What Johansson and Hall don't tell their subjects, until after the experiment ends, is that they secretly swapped the photos on one-fifth of the trials, after each subject made their first choice but before they could expound on why they had made it. Most subjects didn't notice the swap. So instead of explaining why they chose the face they now held in their hands, each subject was in fact explaining why they picked the face they had actually just rejected. And boy oh boy, did they lie.

Johansson and Hall pulled this off by using what magicians call Black Art (similar to that of Omar Pasha in chapter 1), but in this case instead of a black curtain they used a black tablecloth and black-backed photos. In order to fool subjects, they asked them to point to the preferred photo and laid it facedown on the table. That photo had a black back. On top of it they had hidden a second photo, this one of the rejected face. That photo had a red back. When it came time to move the photo toward the subject, the scientists slid the red-backed card (rejected face), leaving behind the black-backed card (preferred face), which was now invisible against the tablecloth. The subjects never saw the swap.

While each subject's brain made up a story for itself to rationalize the "choice," Johansson and Hall (they would take turns serving as experimenter on each sequential subject) surreptitiously swept the actually chosen card off the table and into their laps. Meanwhile the subject assumed that the photo that had been pushed across the table was the same one he or she had chosen. This unspoken assumption served as a powerful method of duplicity.

END OF SPOILER ALERT

The swaps were discovered less than a third of the time. On the successfully swapped trials, the subjects actually confabulated their reasons for having chosen the substitute photo.

One man said, "I preferred this one because I prefer blondes," when in fact he had first chosen a dark-haired woman. One woman chose a woman without earrings, and when the photo was secretly swapped for a woman with earrings, she said she had chosen that one because she liked earrings. Pants on fire! The subjects hadn't chosen the people whose photos they now held in their hands, but *they thought they had*. So what do you do when you are made to justify a choice you believe you made? Confabulate. Stick to your guns.

In a follow-up experiment, shoppers in a supermarket tasted two kinds of jam and then explained their choice while taking further spoonfuls from the "chosen" pot. The pots were rigged so that the subjects effusively praised jam they had previously rejected. A similar experiment was done with tea.

Currently, the researchers have begun to examine choice blindness for moral and political opinion. Using a new tool, a "magical questionnaire," they are able to manipulate people's answers to questions presented in a survey format. Participants are asked to rate to what extent they agreed with a specific moral statement, e.g., "It is morally reprehensible to purchase sexual services even in democratic societies where prostitution is legal and regulated by the government," and then, at the end of the experiment, they are asked to explain why they agreed or disagreed with the statement. Again, the results show that a majority of the participants are blind to the changes made, and that they often construct elaborate arguments supporting the opposite of their initial position.

These studies help us understand how we rationalize many of our decisions. It's not so much the nature of decision making but the repercussions of those decisions that affect our lives.

CHOICE BLINDNESS AS A WAY OF LIFE

Choice blindness works havoc in your everyday life. Have you ever been the victim of the bait-and-switch, where you thought you were buying one thing but came home with something else?

If you truly had free will, advertising and salesmen's pitches would have no effect. For example, when Steve was a postdoctoral fellow splitting his time between two labs, he needed a car to drive between Harvard Medical School in Boston and Cold Spring Harbor Laboratory on Long Island. So he bought a shiny new black Dodge Intrepid ES with a moon roof, motorized leather seats, upgraded rims, Infiniti surround-sound system, and automatic air temperature controls. It was expensive for a postdoc's salary and put a drain on his resources, but he rationalized the decision because it was an incredibly safe car with side air bags (which were new at the time), traction control, an automatic braking system, and other advanced safety features. After all, the long drives between Massachusetts and New York required an extra measure of safety, right?

Sure they did. His decision had nothing to do with thinking that chicks dig a cool car.

> To be fair, he did go to the car dealership with a list of desired safety features. He arrived at the car lot driven by a strong sense of responsibility. The salesperson took one look at Steve's list, knew that the high-end models were the only ones that came with the features he wanted as standard, and then preyed on the fact that his customer was a single male with testosterone-driven needs. Steve could have ordered a cheaper, drabber, smaller model with the same safety equipment and then waited two to three months for the new car to arrive. But the salesman forced him (in the sense that magicians use the word) to buy the fancy car instead.

In the Western world we choose our own mates, right? Arranged marriages and professional matchmakers have joined siegecraft and alchemy in the dustbin of history, have they not? Perhaps. In theory, we can go forth and multiply with anybody we want, so long as there is mutual agreement. We are free, and our number of choices seems for all intents and purposes infinite.

But in practice most of us are no less restricted in our choice of mates than a tradition-bound Eastern youth heading toward an arranged marriage. Consider the fact that we must actually know and interact with the person with whom we pair. We are therefore restricted, in general, to the same geographic location, socioeconomic background, religion, age, current state of availability, and roughly the same level of attractiveness. In reality, it's hard to find a mate who matches all of these parameters, especially after you've completed high school and college. It's hardly a surprise that so many people marry either their high school or college sweethearts.

So how free are we really? Not very. Eastern practices of matchmaking seem fairly intelligent considering that the choices are made by people (usually parents) who care about the couple, who have hard-earned perspectives on the full course of life, careers, and parenthood, and take all of the issues listed above into account during their decision-making process. Further, with notable exceptions in certain isolated parts of the world, the "lovers" nowadays usually have veto power, at the very least.

Finding a great mate (and one whose baggage is lifetime-tolerable—heard any good mother-in-law jokes lately?) requires real luck in the West, and yet it feels completely free. "I make my own luck," say the enlightened, empowered masses. "Believe you will get what you want," says the mega-bestselling self-help book *The Secret*, "and it will manifest." This mass enchantment is one of the grandest magic tricks ever devised.

Why do our choices feel so free and unlimited? One answer lies in a psychological principle called *cognitive dissonance.* This arises when two competing ideas, behaviors, facts, or beliefs are in conflict in your brain. A common way that your brain reconciles the conflict is to change its attitude, beliefs, or behaviors to bring one of the competing ideas into prominence. Magicians love cognitive dissonance, since it leads spectators to feel as though they've made decisions freely for themselves.

An example of this comes from the 2009 Society for Neuroscience annual meeting in Chicago, where we organized a presentation to illustrate the power of magic and its potential usefulness in the lab. Our colleagues Apollo Robbins the Gentleman Thief and the mentalist Eric Mead demonstrated various tricks and magic principles to more than seven thousand neuroscientists gathered in a huge ballroom.

The night before the big event, we saw cognitive dissonance in action when Mead performed a magic trick at a party hosted by the society's president, Tom Carew. Scores of world-famous neuroscientists were gathered in his opulent multiroom hotel suite overlooking Lake Michigan.

At one point, Mead had a scientist pick a card from a deck and then asked him to randomly spread out all the cards over a large area of the floor. Only the scientist knew which one was the chosen card. Then Mead took one end of a linen napkin, handed the other end to the scientist, and, pulling it tight, dragged the fellow around the strewn-out cards. Mead boasted that he would detect minute changes in the napkin's tension and thereby read unconscious signals from the scientist's mind to find the correct card. After a minute of this performance, Mead found the card.

The interesting thing about this trick* is that after the party, when people were tittering to each other during the descent to street level in the elevator, the scientist who participated in the trick opined that Mead must have known in advance which card he would choose. This was met by a quick denial from another scientist, a world expert in the field of motor control, who said it was no trick at all. To her mind, Mead had clearly used neuromuscular feedback from the napkin to find the card. She knew that Mead had made no bones that it was a trick, and yet here she was, arguing for something far less likely. Swept up in the moment, her cognitive dissonance had taken her for a wonderful ride down a magical road.

When you make a decision between two things that seem equivalent, cognitive dissonance frequently comes into play. You elevate the value of your choice for the simple reason that it was your choice. Have you ever had a boss who made a dumb decision that became immutable policy long after she realized she had been in the wrong? Cognitive dissonance. Have you yourself ever made a dumb decision concerning your children, but then stuck to your guns so as to "provide consistency"? Cognitive dissonance. Have you ever looked down on people who live in a rival sports team's city for no other reason than that their zip code places them in the enemy camp? Cognitive dissonance.

Cognitive dissonance happens because our free will isn't truly free; it's highly constrained by our context and history. And history, we know, is written by the victors. This is as true of the potential thoughts and deeds that populate our minds as it is of cultures and nations: the winning choice orchestrates emotion, language, and memory to make itself the inevitable and infallibly correct one. In reality, all behavioral decisions are nothing more than a reflection of our genetic and environmental history.

Many people get upset when neuroscientists and philosophers state that free will is an illusion. Those who believe that the mind is wholly

*We cannot reveal exactly how this trick was accomplished, because Eric Mead has asked us not to tell his secrets. Suffice it to say, it was a trick.

separate from the brain—a supposition called *dualism*—tend to believe that free will is a fundamental property of the mind. According to this view, free will is a separate, numinous quality of being that is not subject to physical laws or reducible to chemistry and circuitry.

But in the realm of neuroscience, there is not a shred of evidence for dualism. The mind is what the brain does. Consciousness and mind are products of your brain.

How could that be? You feel as if you are in full control of your mind. Sure, your brain carries out many tasks without your being conscious of them. You drive home on automatic pilot. You put cups into a dishwasher while carrying on an interesting conversation. But making important decisions? Isn't mental life dependent on the fact that you are free to choose among different possible courses of action? Your decision-making process seems to be driven by your own volition. This feeling fits your sense of justice and moral responsibility.

Let's look at several lines of evidence for the idea (dare we say *fact*) that free will is an illusion. In the 1970s, Benjamin Libet, a neurophysiologist at the University of California, San Francisco, carried out a series of studies that first challenged the notion that we make decisions consciously and freely. Libet asked people to stare at a clocklike timer with a ball moving around the periphery once every three seconds. They had to press a button with their right index finger whenever they felt an urge to do so and afterward tell Libet where the ball was (what time it was) when they decided to make the move. Two testing devices—an EEG (electroencephalograph) and an EMG (electromyograph)—recorded their brain activity and the electrical activity of their muscles. Libet found that participants had the conscious sense of willing the movement about 300 milliseconds *after* the onset of the muscle activity. Moreover, the EEG showed that neurons in the part of their motor cortex where movements are planned became active a full second before any movement could be measured. You might think that the delay was due to the conduction time between the brain and the muscles. But a full second? No way. There was definitely something interesting happening here.

The findings mean that your brain unconsciously makes the decision to move well before you become aware of it. In other words, your brain, not your conscious mind, makes the decision. This does not match your experience, but it is how your mind actually works. Before you get discombobulated, know that there is a silver lining to these results: while decisions are unconsciously prepared ahead of time, you can still veto your actions. According to Libet and others, you may not have free will, but you do have "free won't."

UNWILLINGLY WELL ENDOWED

The feeling of free will is pervasive to our psyche, but carefully designed laboratory conditions, such as in Libet's experiment, can reveal free will for what it is: a sophisticated cognitive illusion. And if we pay close attention, we can also find rare instances in our everyday life in which the illusion breaks down. Have you ever been flung uncontrollably down a trajectory of complex behavior that you couldn't control no matter how hard you tried? We're not talking about bodily functions like coughing, sneezing, or orgasm. Those are certainly complex behaviors in which you feel dissociated from the actions of your body, but they are reflexes rather than choices. Drug addicts, alcoholics, and patients with a variety of neurological disorders lose the sense of free will, but what about healthy people?

We saw a great example of someone "losing it" in 2005 while visiting Susana's hometown of A Coruña, Spain. A politician in the mayor's office, Carlos González-Garcés, was on television giving a boring press conference on a new program regarding the city's fire safety readiness.

"This last year twelve specialization courses were offered, with a very strong focus in the courses given to new firefighters," he said. Some minor details about firefighter courses followed before he began discussing the fire department's equipment status.

"They are well endowed," González-Garcés told the reporters. He gave a small smile and corrected himself: "They are well endowed in regards to material resources." But then he realized he'd made the situation worse and gave an even bigger smile, which he then tried to suppress. The poor guy tried to hide his face by looking down and to the side of the bank of microphones. "As a matter of fact, this year a concrete investment was made." He was openly laughing now, punctuated by giggles. "I did it by accident," he said, and began rocking side to side, as if to stave off peeing his pants. By this time, his staff was laughing with him. "It was not premeditated," he assured the press. González-Garcés regained control briefly but then lost it as the reporters' guffaws could be heard in the background. "Okay, let's see," he said before another failed attempt to suppress laughter. He was now wiping the tears from his eyes. "Never . . . this never happened to me before." He wiped his eyes again and tried to plow ahead. "All right, so the thing is . . . a specific truck was bought . . ." but he couldn't keep himself from another fit of laughter ". . . for the old part of the city." He was giggling again. "Ay-ay-ay." He sniffled against his running nose and once again failed to suppress the laughter. Like a marionette on a string, González-Garcés threw himself against the back of his chair, convulsing with mirth. "This seems so childish . . . it is a laugh attack." He snorted, wiping both eyes. "Ay-ay-ay. Okay. Forgive me." He cleared his voice, brought his chair closer to the table, sniffled, cleared his voice again, and suppressed his giggles. "Okay, here you have the number of vehicles," he started, but he was still completely out of control. He threw himself back in his seat again, laughing uproariously. "And no further explanation is needed," he explained by way of surrender. "If you have any questions about the firemen's endowment," he added between guffaws, "the gentleman who is in charge can answer you."*

The politician's uncontrolled laughter was not a reflex, which by definition is a process that takes place in the shortest possible route through a given neural pathway. When the doctor hits your knee with a hammer and your leg jerks, that is a reflex. No brain required. Laughter, on the other hand, involves a highly complex series of emotional, cognitive, and motor actions that you think you can control. You always have the option of not laughing when you don't want to, right? You have control over your body and behavior, correct? Wrong. This example of the poor guy laughing so hard he almost wet himself on TV shows that while we feel we are in control, we are actually just along for the ride.

*See http://sleightsofmind.com/media/laughattack.

A colleague of ours, John-Dylan Haynes of the Max Planck Institute in Berlin, Germany, recently reprised Libet's work using functional brain imaging. He wanted to see what happens in people's brains when they make conscious choices. If you had taken part in the study, you would be lying in a scanner when Haynes tells you that you can decide if you want to press a button with your right hand or left hand. You are free to make this decision whenever you want, but you have to remember the time when you feel you have made up your mind. The researchers used a sophisticated computer program trained to recognize typical brain activity patterns preceding each of the two choices.

Haynes was astonished to find that brain signals—tiny patterns of activity in your frontal lobes—predict your decision (that is, whether you will press the button with your left or right hand) up to *seven seconds* before you make a conscious choice. This means that parts of your brain can sometimes know what choices you are going to make several seconds before you become consciously aware of them. Because these brain areas are clearly active with information indicating the choice you are about to make, well ahead of the time that you consciously feel you've made a decision, it seems likely that these brain areas serve to bias your upcoming decision. You may be convinced that your decision was a free, open choice, but it's just not true.

If your actions are determined by the prior neural activity happening in your unconscious brain seconds before you consciously make a decision, do you have a choice about anything? Are you responsible for what you do? In his book *The Illusion of Conscious Free Will*, the Harvard psychologist Daniel Wegner digs into such questions by comparing the illusion of free will to the perception of magic. You perceive magic, he says, when an apparent causal sequence (the magician saws his assistant in half) obscures a real causal sequence (the box is rigged so the saw blade never touches her). You do not perceive the real thing even though the apparent sequence violates common sense and you know it's impossible.

Wegner argues that the "self" is magical in this same sense: "When we look at ourselves, we perceive a simple and often astonishing apparent causal sequence—I thought of it and it happened—when the real causal sequence underlying our behavior is complex, multi-threaded, and unknown to us as it happens."

Wegner wonders how people develop this magic sense, what the philosopher Daniel Dennett calls "some concentrated internal lump of specialness." Why do we experience our actions as freely willed, arising mysteriously from the self? And why, too, do we resist attempts to explain those actions in terms of real causal sequences, events that are going on behind the curtain of our minds?

We feel as if we have free will because we have independent thoughts and desires that are then acted upon accurately by our bodies. Our brains are correlation machines, as the magicians prove to us over and over with the presentation of impossible causal events. Because we have the ability to connect cause and effect, there is no evolutionary pressure to develop the sensory pathways necessary to track every bit of the information flowing through our brains. Remember that our neural resources are limited and that we cannot attend to everything in our visual field. Well, that attentional limit would be even more woefully deficient if we also had to attend to every single little process in our brains. Do you really want to know every minute detail of the information that the neurons in your prefrontal cortex are sending to your primary motor cortex in order to reach for a glass of water? Suffice it that when we are thirsty, our arm successfully picks up a glass of water and brings it to our mouth. We conclude that our free will directs the action because we didn't tell anybody else about our internal wishes.

Wegner designed an experiment to see if he could prime people to experience thoughts consistent with an event they did not cause and if they could be convinced that they caused it.

Roll back the clock and assume you are a participant. You are asked to help with a study on psychosomatic influences on health. Your task is to play the role of a witch doctor who lays a voodoo curse on another participant, a victim, by sticking pins into a doll. In reality, this person is a confederate in the study (she works for Wegner). Not long after you jab the pins into her ersatz doll body, she feigns a headache. Would you believe you caused her headache? Many of the study participants did. Moreover, if the "victim" acted in an obnoxious manner, the level of magical witch doctor thinking increased. But no harm had really been done at all.

This readiness to make correlations illustrates the general processes by which people succumb to the belief in the paranormal, especially clairvoyance, precognition, and psychokinesis, says Wegner. Our bodies respond effortlessly to our wishes, and we witness the result as a correlation between our wishes and our body's reaction. It's not too far afield, then, for us to wish for the improbable and, when it happens, to believe that we caused it with our hopes and prayers.

Because we are so used to getting what we wish for in life (like one foot stepping in front of the other), we can't stop ourselves from wishing for the physically forbidden. An exception may be the ancient Greeks, who believed that each of their motivations and feelings was granted to them by a god. Chuck Palahniuk, the American novelist, explains, "Apollo was telling them to be brave. Athena was telling them to fall in love. Now people hear a commercial for sour cream potato chips and rush out to buy, but now they call this free will. At least the ancient Greeks were being honest."

You can prove to yourself easily enough that the universe does not accede to your every whim. Wish to perform a Chopin étude on the piano when you've never taken a lesson, and it won't happen. But Wegner explains why we nevertheless overextend our propensity for wishful thinking: "If our wishes seem to prompt a range of activity within our personal sphere of influence, why not hope for more? Many forms of supernatural belief, including belief in prayer, may develop as a natural next step from the magic we perceive in ourselves. If mere wishing can pop the lid off a bottle of beer, why not wish for the moon?"

Two psychological effects further influence the illusion of free will. In the *priority effect*, your sense of agency seems causal when the thought of an action occurs just prior to the action. For example, you can be led to experience the arm movements of another person as if the movements were your own. In our professional opinions as neurobiologists, we can tell you that this effect is downright freaky. Imagine you are draped in a robe, arms at your sides. A helper stands behind you and puts his arms through the sleeves. He wears gloves.

You hear instructions for how to move your arms through a headset. As the helper makes the movements, you feel as if you have control over his arms. It is an illusion of agency. Has anyone ever called you on the telephone at the same time you were thinking about him or her? It's a coincidence, but you feel agency. But then every feeling of free will that you have is an illusion of agency.

In the *exclusivity effect*, you perceive that your thoughts cause events for which there are no other plausible explanations. But there may be reasons for making choices that you are not aware of. Wegner gives a nice example. Say you are at a restaurant and the person next to you orders the shrimp special. You were about to order that but, wait, it might look like you were copying that person. So you change your order so as not to look influenced by the other. You think you are choosing of your free will but it isn't so. The fact that you can be influenced about something as trivial as an order of shrimp shows that your free will is a wet tissue. Indeed, no idea is an island.

Wegner says that we have only our conscious thought and our conscious perception to explain our actions post hoc. We may believe that they are connected to free will, but when we do so we take a mental leap over the demonstrable power of the unconscious that guides our actions and conclude that the conscious mind is the sole player. Your conscious thoughts merely provide a rationale for what you just did, which was motivated in a very unfree, deliberate way by your unconscious brain.

Can you break the spell? Some worry that if we prove free will is an illusion that arises from the flesh, the human spirit will be dead. But such a shift in popular thinking is not likely to happen. The ubiquity of perceived conscious agency in our everyday life is sufficient to quell our inner skepticism telling us that our behaviors are caused by brain mechanisms and not by our free will. The illusion of the magic self cannot be easily suppressed. Moreover, many philosophers and scientists argue that conscious will may be an illusion, but responsible, moral action is quite real.

CAN A MACHINE READ YOUR THOUGHTS?

Can a machine read your thoughts? Can scientists read the contents of your mind via functional magnetic resonance imaging?

The answer depends on what you mean by "thoughts." Functional magnetic resonance imaging, or fMRI, has come a long way since its discovery in the early 1990s. In a nutshell, the technique measures brain activity by tracking increased blood flow, the idea being that more active brain regions will use more energy and will "light up" in the scanner. In the early days of fMRI research, scientists located regions that specialize in things like our basic sensory processes, speaking, reading, or feeling strong emotions. More recently, they found areas specialized to recognize faces or places.

But can the machines reveal what you are thinking? At the MRC Cognition and Brain Sciences Unit in Cambridge, England, scientists are using a new computational technique called *multivariate analysis* to predict your thoughts based on observed patterns of activity. If you were in their scanner, they might ask you to imagine playing tennis and then to imagine walking around the rooms in your home. Based on the patterns observed, they could tell you which activity you were thinking about.

Thus far such studies are highly constrained. Only a handful of mental states have been correlated with brain patterns, which are noisy, indirect measures of neural activity. For example, you could imagine playing soccer and moving around your office, and the machine might not be able to tell the difference. Thus researchers cannot do genuine mind reading—they cannot tell you that you are thinking of a hippopotamus, silently reciting the Gettysburg address, or wondering what you will have for dinner tonight. Mind reading remains science fiction.

10

WHY MAGIC WANDS WORK

Illusory Correlations, Superstition, Hypnosis, and Flimflam

In the winter of 1983, Susana sat at a table in her mother's dining room with her friend Beatriz and her sister, Carolina. The three adolescent girls leaned over a Ouija board that Susana had made the week before. A Ouija board, in case you've never used one, is a device that allows players to ask questions of the spirit world and find answers from a marker, called a planchette, that moves, apparently of its own accord, across a board marked with letters of the alphabet, numbers, and the words "yes," "no," "maybe," and "good-bye." Susana was giddy with anticipation. Unlike her sister and mother, she was the cold-blooded skeptic in her family. This would be fun.

The girls placed their fingertips gently on the planchette, which—instead of the usual heart-shaped device perched on three legs—was a huge silver coin embossed with the face of Spain's deceased dictator, Franco. A radio was blaring as Susana's mother shook her head in disapproval and walked away. To bug her little sister, Susana asked, "What is the name of the boy that Carolina is in love with?" Carolina scoffed and rolled her eyes.

Then the Ouija board started to work. The planchette moved, as if on its own, to the letter *J*. After a brief moment, it began its journey to the letter *A*. Susana thought back to earlier that day when she had told Mother Silvia, one of the nuns at school, that she had made her own Ouija board at home. The nun urged Susana not to use it. "You can't be sure who you're talking to," the nun said. For many Christians, the Ouija board is a gateway to satanic control or demonic possession. For them, channeling or consulting the spirits of the dead is a serious sin.

Now the planchette moved toward the letter *V*. Suddenly a lightbulb exploded in the chandelier above the table. The girls shrieked and pulled their hands back from the sinister device, shaking with an odd mixture of terror and delight. What had just happened? Was the exploding lightbulb a coincidence? Or a warning from the spirit world? Was the explosion and imminent spelling of "Javier" a genuine correlation or an illusory one?

Susana recalls that she and the other girls were all a bit freaked out but pleased to have gotten such a strong "manifestation" before stopping the game for the day. Susana's mother was fairly superstitious, the type of person who believes that you shouldn't tempt fortune. She wouldn't say the word "snake" or "viper" because that would bring bad luck. Susana's mother and sister believed that a supernatural hand had moved the planchette—and smashed the lightbulb.

As cognitive neuroscientists who study foibles of the human mind, we see superstitious beliefs as examples of illusory correlation, the phenomenon of seeing a relationship between events when there is no factual evidence of such a relationship. Exploited by magicians and psychics alike, illusory correlations are the basis of stage acts, magical thinking, and all manner of flimflam. They also can cause enormous mischief in this world.

As scientists, we can explain how Ouija boards work. Spirits don't move the planchette; you and the other players do, via what is called the *ideomotor effect*. Your voluntary muscles can make tiny movements outside of your conscious awareness. When the movements of all the players reach a consensus (again, unconsciously), the planchette drifts toward a letter, then another, and so on. The ideomotor effect explains other supernatural phenomena including dowsing, automatic writing, and facilitated communication. The movements are self-generated, yet the illusion of an outside force is compelling.

You can try the ideomotor effect on yourself. Suspend a handheld pendulum over a sheet of paper on which you have written the words "yes," "no," "maybe." Ask any and all sorts of questions and the pendulum will give you answers.

And if you want to expose the illusion of the Ouija board, ask the players to put on blindfolds as they move the planchette. Their spelled-out messages will be gibberish.

Teller, the mute partner of the famous Penn & Teller duo and master of illusory correlation, walks on stage at the Magic of Consciousness symposium.* He is small—but taller than you'd guess from seeing him next to the imposing six feet six of Penn Jillette on television or in their stage show at the Rio Hotel and Casino in Las Vegas—and agile, with an impish personality. Dressed in black trousers, black loafers, and a black shirt with dragons stenciled on the back, he looks like an elf king from the *Lord of the Rings.* Teller almost always wears a bemused expression, as if concealing a private joke, and clearly relishes the opportunity of explaining magic to the assembled scientists.

"One of the things magicians do," says Teller, "is take advantage of our natural inclination to study something we see done over and over again and think that we're learning something. Because in real life, if you see something done again and again, you study it, and gradually you pick up a pattern. If you do that with a magician, it's sometimes a big mistake."

Teller paces the stage. "Suppose I produce a coin." He holds his right hand high over his head and out of thin air produces a shiny silver coin. Then Teller drops the coin into a brass bucket held in his left hand. You hear a loud clink. He thrusts his right hand in a different direction and snatches another coin from the air. *Clink.* Then another. *Clink.* And another. He bites it and says "This one is real" before dropping it into the bucket. *Clink.* "Maybe another one out

*See http://sleightsofmind.com/media/magicsymposium/Teller.

there." *Clink.* With this last coin, Teller spreads his hand and fingers wide so that you can see he is not hiding anything.

Then Teller walks up to the audience and, combing his fingers through a man's white hair, pulls out yet another coin and tosses it into the bucket. *Clink.* He removes a guy's eyeglasses and tips the lenses over the bucket's lid. *Clink clink.* He rapidly picks up another person's bag, rummages through it, and pulls out more coins. *Clink clink clink.* Finally he holds his hand up to his face and coughs, and out falls yet another coin that goes into the bucket with the rest. *Clink!*

The effect, called the Miser's Dream, is a classic. It goes back at least as far as 1852, to the magician Jean-Eugène Robert-Houdin, who called it the Shower of Money. Later it became known as the Aerial Treasury, until in 1895 T. Nelson Down named his version the Miser's Dream, a name that has stuck. For the trick, the magician continually plucks coins out of the air, seemingly from anywhere he wants, and tosses them into a receptacle. The act is traditionally done in silence except for the loud clanking of coins accumulating in the receptacle.

"Your natural inclination as an observer is to assume I am doing the same thing over and over again," says Teller. "Now I will tell you exactly what I did so you can see how devious we are."

SPOILER ALERT! THE FOLLOWING SECTION DESCRIBES MAGIC SECRETS AND THEIR BRAIN MECHANISMS!

Teller explains that he began by palming five coins in his right hand. His left hand holds six more coins that are pinned with his fingers against the inside wall of the bucket. Some coins will drop from his right hand, and some from his left (which you cannot see) while he pretends to drop a coin from his right. In the latter case he is only faking the action of tossing from the right hand, using the flick-down motion to reconceal the coin. But the faked action engages your mirror neurons, so you are predisposed to see it as the same natural tossing action you yourself perform daily with coins, car keys, cooking ingredients, and so on. The clink of the coin dropping into the

Teller relies on misdirection and sleight of hand to create an illusion called the Miser's Dream. (Photographs © Misha Gravenor)

bucket from the left hand helps create the illusion that the fake-tossed coin from the right hand landed in the bucket. What we're actually seeing is the same coin flash in the right hand over and over and over again. Your assumptions have misled you.

Teller says that the first coin raises the question in your mind: Where is it coming from? After four coins, you think you know. He has to be palming them in his right hand. Just then Teller reveals that his right hand is completely empty except for a single coin held between his thumb and index finger. You conclude there are no hidden coins. But wait. He is still dropping them, *clink clink*, into the bucket, only now they are coming from his left hand. "Every time you think you know what is happening, I am changing the method," he says. Every coin is a new little burst of sight and sound—you see it, you hear it, and it is all happening so fast you are deceived. You think any repetition is a real repetition.

END OF SPOILER ALERT

Teller continues, "Your natural inclination as an observer is to assume that what I'm doing is the same thing over and over again. We take for granted that a repetition is a repetition [even] when it's not."

"We all infer cause and effect in everyday life," adds Teller. When A precedes B, we conclude that A causes B. The skilled magician takes advantage of this inference by making sure that A (a fake coin toss) always precedes B (a loud clink). However, A does not really cause B.

Teller's performance of the Miser's Dream reveals the human compulsion to find patterns in the world and to impose them even when they are not actually there. The magician milks your instinct to infer cause-and-effect relationships. This is similar to how magicians use your own expectations against you (as we discussed in chapter 8). But here we're talking about how magicians make you see correlations that aren't really there. They hijack your powerful abilities to detect patterns in the natural world, and then they trick you into drawing correlations between the unexpected, the ridiculous, and the absurd. They then pin your cognitive processes to the floor like a bully sitting on your chest as you wrestle with the contradictions your own mind conjured up.

As you saw with the Ouija board incident, this is the *illusory correlation* effect. In most circumstances, our inborn instinct for inferring cause-and-effect relationships serves us well. Want an egg? Look in a bird's nest. Dark clouds gathering overhead? Rain is likely, go find some shelter. That's all well and good, but causal inference is a highly imperfect, eminently fallible faculty. It goes amiss all the time and leads us to believe all kinds of things.

Illusory correlation is at the root of why some people honestly and in all good faith believe they are psychic. The telephone rings and you were thinking about the caller at that very moment. You sit down at your computer to write an e-mail to a friend only to discover that your friend has just written you about the same subject. You may know someone who believes he has predicted the future in a dream—a plane crash, say. But what he doesn't tell you is that he has premonitions of a plane crash several times a week. He tends not to notice or remember these false predictions, but the one that coincides with an actual plane crash sets off wild alert bells in his brain. His mental correlation detector is screaming *Correct! True! Valid!* In extreme cases, illusory correlation can lead to extraordinary beliefs, such as the ancient Aztec theory that a human sacrifice had to be performed each morning in order to make the sun rise. It's gruesome, and easy to condemn in hindsight, but to the Aztecs it worked every single morning, just as advertised.

In the second season of the television series *Lost*, plane crash survivors stranded on an island must push a mysterious button every 108 minutes to "save the world" (prevent the occurrence of some undefined world-scale catastrophic event). Because the end of the world has not yet come, the button must be working. But nobody ever fails to push the button to find out for sure.

A related effect in the brain is called the *availability bias*. This illusion, caused by a failure of memory, pops up often in everyday life. For example, according to Steve, "I change our baby's diaper *waaaay* more than Susana does. Evidently because she's lazier than I am." But the puzzling thing is that Susana thinks exactly the opposite. She thinks she changes Brais's diapers more than Steve does. The fact is both of us are wrong. We each change Brais's diaper more or less equally often. But in our minds, our own contributions and sacrifices are magnified by the fact that we remember our own actions better than we remember those of others. We incorrectly draw stronger correlations between the facts that we remember than between facts that are provided by a third party.

Magicians are well aware of these little brain foibles, and they pump them like a lab rat on a cocaine lever. "Much of our life is devoted to understanding cause and effect," Teller says. "Magic provides a playground for those rational skills. It is the theatrical linking of a cause with an effect that has no basis in physical reality but that, in our hearts, ought to. It is rather like a joke. There is a logical, even if nonsensical, progression to it. When the climax of a trick is reached, there is a little explosion of shivery pleasure when what we see collides with what we know about physical reality."

This "little explosion of shivery pleasure" can actually be studied in the laboratory. In 2009 a team of cognitive neuroscientists—led by Ben A. Parris and Gustav Kuhn of the universities of Exeter and Durham in England—used magic tricks to investigate the neural correlates of causal relationships etched in the brain through experience. In their study, they point out that a magician screws with your head when he puts a coin into his right hand, closes it, waves his left hand over his clenched fist, and then slowly opens his right hand. The coin, which you know must still be there, has vanished. Your implicit system of knowledge of cause and effect tells you that coins cannot disappear like that.

So what goes on in the brains of people who witness such tricks? To find out, the researchers scanned the brains of twenty-five people with fMRI as they watched video clips of various magic tricks and two closely related control conditions. For example, a trick might be like the one just mentioned: coin in hand, close hand, wave other hand, open hand where coin last seen, coin gone. A control condition would be: coin in hand, close hand, wave other hand, open hand where coin last seen, coin still there. A surprise condition would be: coin in hand, close hand, open hand, magician shows coin in his mouth.

The main finding was that two brain regions—with the mouthful names *dorsolateral prefrontal cortex* (dlPFC) and *left anterior cingulate cortex* (ACC)—lit up when people watched the magic tricks. Research has shown that one of these areas, the ACC, detects conflict, whereas the other, the dlPFC, tries to resolve conflict—exactly what you would expect when a cause-and-effect relationship is violated. In the surprise condition, the conflict detecting area, the ACC, lit up along with another region of the prefrontal cortex called the *ventrolateral strip*, which has been found to register surprise. In the plain vanilla control condition, none of these areas showed increased activity. The researchers concluded that your ability to detect information that contradicts or challenges your established beliefs is crucial for learning about the world. The highlighted circuit seems to play a role in the neurobiology of disbelief.

Susana's sister, Carolina, all grown up into a slim, chestnut-haired beauty, is a supervising croupier in the casino of León, Spain. She has seen more than her fair share of customers whose thinking is dominated by a peculiar cognitive illusion beloved by magicians and charlatans the world over: the *gambler's fallacy*.

"Clients often ask, how long has it been since the number twenty came up?" says Carolina. "Well, we croupiers keep track of every spin of the wheel, and since there is no rule against it, we truthfully answer ninety-six balls ago." And why should it be against the rules? It plays in the house's favor that customers are drawn along by the illusion that knowing the past will help them predict the future. Carolina explains that modern roulette wheels come outfitted with electronic counters that conveniently provide various statistics for the gamblers' "benefit," such as the numbers corresponding to the last fifteen balls, the percentage of black versus red numbers, the "hot" or most frequent numbers, or the more frequent dozens (numbers 1 through 12, numbers 13 through 24, or numbers 25 through 36). Of course none of these statistics changes the fact that the ball has exactly 1 in 36 chances of landing on any given number on the next spin.* It should come as no surprise that Carolina, like many croupiers, doesn't herself gamble.

The gambler's fallacy is the mistaken belief that the likelihood of an event increases when a long period has elapsed since the event last occurred. If you're in a drought, it feels more likely that it should rain tomorrow. If you and your spouse have had four daughters in a row, it feels likely that you'll have a boy the next time. And when you're gambling, if it's been a very long time since the ball landed on 20 on the roulette wheel, it feels as if the likelihood of an impending 20 is high.

One of the most memorable examples of the gambler's fallacy took place at the ornate Monte Carlo casino in 1913. Elegantly dressed gamblers stood around a roulette wheel and watched as the ball landed on black twenty-six times in a row. With increasing excitement, many patrons began betting on red. It just *had* to come up next. Sure, the wheel is random, but it had to "self-correct," right?

Wrong. We all succumb to the superstition that when we observe a random process with a deviation, then logically the imbalance will have to even itself out. For example, ask yourself, if you toss a coin seven times, which is more likely to be the result? Heads, heads, heads, heads, heads, heads, heads. Or tails, tails, tails, tails, tails, tails, tails. Or heads, tails, tails, heads, tails, heads, heads.

*Contrary to the gambler's fallacy, a roulette number that has occurred more often in the past may be more (rather than less) likely to recur in the future. The reason is that no roulette wheel is perfectly manufactured. Real-life wheels are sometimes slightly biased, and they will have very small tendencies to land more on certain numbers. In the early 1990s, the Pelayos, a family of Spanish gamblers, secretly recorded roulette results for extended periods of time in Spanish, Dutch, and even Las Vegas casinos, and they successfully predicted that some numbers were slightly more likely than others to come up. They quickly amassed a small fortune, and just as quickly were banned from further casino play.

Answer: they are all the same. Each is an independent fair toss. The coin has no memory. If you toss twenty tails, the probability of flipping another tails is one in two. You can choose the same lottery numbers every time or change them every time, but either way you are equally likely to win an individual lottery draw. You could use the numbers that won the previous day and have the same probability of winning. The universe does not carry a memory of past results that will favor or disfavor future outcomes.

TWO GOATS AND A CAR

In September 1990, the "Ask Marilyn" advice column in *Parade* magazine posed the following puzzle. Suppose you are in a game show and you're given the choice of three doors. Behind one door is a car; behind the others, goats. You pick a door—say, number one—and the host, who knows what's behind all the doors, opens another door—say number three—which he knows conceals a goat. You look at the goat as he says to you, "Do you want to stick with door number one or switch to door number two?" What should you do? Is it to your advantage to switch your choice?

The puzzle, known as the Monty Hall problem after the host of the popular American television game show *Let's Make a Deal*, tests your ability to assess probabilities. You don't know which of the two remaining doors hides the prize, and so you may think, hey, the odds are fifty-fifty. It feels right to stick with door number one. But you'd be wrong. According to experts on probability, you should always switch. Choosing door two doubles the probability of winning the car from one-third to two-thirds. The Monty Hall problem arises because the contestant correctly believes that there is a 1 in 3 chance of selecting the car door in the initial door choice. But the host then removes a goat door from the remaining two doors. Now, if the contestant did indeed choose a car door in the original round (a 1 in 3 chance), then the remaining door will contain a goat. But if the contestant chose a goat door in the original round (a 2 in 3 chance), then the remaining door will contain the car. So it's twice as likely that the contestant's original choice was a goat rather than a car, and since it is certain that one of the remaining doors must hide the car, it is always in the contestant's best interest to switch.

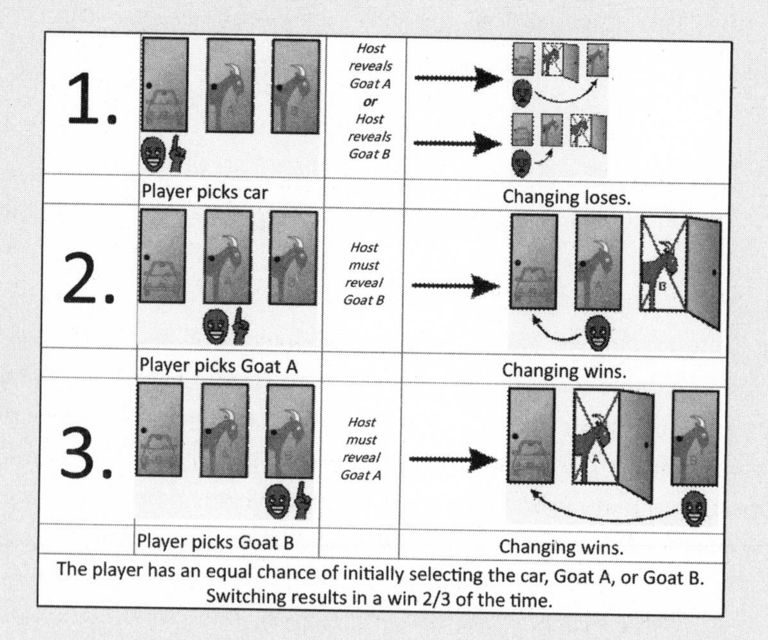

The trouble is, the solution doesn't feel right. It doesn't match your intuition. And you are not alone. When the puzzle was first published, many scientists, including one or two Nobel Prize winners, were outraged by the correct solution because it did not feel right to them, either. Equal probability is deeply rooted in intuition.

The gambler's fallacy may manifest when a gambler suspects that a roulette wheel is rigged. If no red shows up after a long string of blacks, the player may assume that the wheel is not on the up and up. Someone must be cheating. Mentalists have taken this observation to heart and devised what they call the "too perfect" theory in magic. When producing a series of continuous predictions (for example, divining what's written on a bunch of notecards collected from the audience), magicians will often elect to get a few predictions wrong. They reason that psychic ability should be imperfect. After all, if the magician or psychic gets every prediction exactly right, the act ceases to look supernatural. If the mentalist never misses, the audience assumes the trick is "rigged" and not psychic.

WHAT ARE THE ODDS?

In 1937, Susana's grandfather Enrique García Casal, then twenty-two years old, was conscripted into the Spanish Civil War, an epic three-year struggle between an army led by General Francisco Franco and the democratically elected government, the Second Spanish Republic. During the last few days of the war, Enrique found himself aboard an armada headed for Cartagena, a Mediterranean coastal paradise, and one of the last Republican strongholds. He and his fellow soldiers had been told that Cartagena just surrendered. The war was nearly over and their job would be to occupy the defeated city.

Unfortunately for them, the fleeing Republican army had maintained control of the Cartagena coastal batteries and Enrique's ship, the *Castillo de Olite,* was in its sights. Franco's navy called for the armada's retreat but the *Castillo de Olite*'s radio was broken and it continued toward the harbor, fully intent on landing its troops.

A floatplane circled and waved. "What a wonderful welcome surprise!" Enrique thought. In actuality, the floatplane was sent by the Republicans as a last-ditch effort to warn off the boat. Remember, this was a civil war: nobody wanted to kill a massive ship full of sitting ducks when they were fellow countrymen. Why there might be family members on board!

Finally, the leader of the coastal batteries ordered that the ship be sunk. When the guns first fired, Enrique and his fellow soldiers rejoiced at the impressive welcoming salute. So it came as a big surprise when a round from the huge Vickers guns hit the water nearby. A giant plume of water and foam sprayed into the air.

Enrique was at the stern of the ship when a round hit the bow. A second explosion sent debris and body parts flying in every direction. Crew and soldiers began to abandon ship. Enrique looked down to discover he was grazed at waist height with shrapnel. He bled as he considered that, had he been standing two inches to the side, he probably would have been a goner.

Half crazed, Enrique took off running and swan-dived off the boat. He was a strong swimmer and moved quickly away. Of the 636 men who made it off the sinking ship, Enrique was one of the last to jump.

And that's when the situation went from horrible to absurd. Enrique's cousin from his hometown of A Coruña, a city in Spain's far northwest, stopped Enrique in the middle of the Mediterranean and pointed out that he was swimming toward Africa, rather than the closer shores of Spain. Enrique was astonished to see his close relative. They turned, swam to a small island in the Cartagena harbor, and were rescued by the lighthouse keeper and his wife. The Republican army captured them and held them prisoner for the short remainder of the war.

Susana's family tells this story as if it were evidence of divine intervention. What, they ask themselves, are the chances that you are party to the greatest maritime disaster in Spanish history, on a boat with thousands of random soldiers, and you happen to run into (well, swim into) your cousin, who you didn't even know was on board? Surely the probability is small. But is it as minuscule as it seems?

Consider that Enrique was one of eight children and his father also had seven siblings. Enrique had dozens of first and second cousins, all about the same age, living in or around the same city. Half of these cousins were men and many would have been conscripted at the same time. Moreover, the military commonly drafted troops by the truckload and kept them more or less grouped in their units according to neighborhood. Indeed, the vast majority of the men on Enrique's ship were from the same part of Spain, so the likelihood that he would run into a cousin in the water could have been as high as 10 percent.

It is extremely easy to miscalculate probabilities and to assign inordinate significance to merely unlikely events. In magic shows, mentalists are masters at promoting unlikely events to the point that they seem impossible. In this way, only magic or some other divine intervention can seemingly explain the effect, when, in fact, if you actually inspect the series of small coincidences that led to the outcome, it's not so very surprising.

Of all the ways you can be suckered in by the supernatural, putting your faith in a psychic arguably tops the list. Mind reading as performed by magicians is one thing; they have mastered elaborate tricks that allow them to be in full control of events. In their "hot readings" (mentioned in chapter 7) they learn as much as they can about you before the show by trawling the Internet or government records, overhearing conversations, or even lifting your wallet for a quick peek. As we saw in chapter 9, they also trick you into picking specific words or numbers that feel like free choices. Armed with this knowledge, they appear to read your mind by regurgitating what they know about you.

Psychics, on the other hand, are not magicians. Although they may occasionally rely on hot readings, they are fundamentally masters of so-called cold readings, which are not meant to entertain you but to gain your trust and, all too often, defraud you. In a cold reading a magician, mentalist, or psychic draws information out of you to give you the impression that he is reading your mind. The method relies on an ability to sense unconscious behavior and to spin out vague statements that fit anyone's situation. The deception is all linguistic. There is nothing supernatural about it.

Nevertheless, we wondered whether psychics might have anything to teach us about the neuroscience of human behavior. Even if what they do is all trickery and hokum, maybe psychics are fundamentally geniuses of the mind, like magicians and mentalists, and we should be studying them, too, to improve neuroscience.

In April 2010, we dug out our tie-dyed T-shirts and headed off to Sedona, Arizona, to attend a psychic fair. We could feel the "positive energy" as we approached beautiful red rock formations eroded from the iron-rich landscape. Psychics, faith healers, and New Age entrepreneurs populate the area and assert that Sedona is one of fourteen power points on earth that can "ground the vibrational frequencies" coming in from extraterrestrial sources. (The other hot spots are Haleakala in Hawaii, Mount Shasta and the Golden Gate Bridge in California, the Black Hills of South Dakota, Central Park in New York City, Machu Picchu in Peru, Mount Olympus and Delphi in Greece, Japan's Mount Fuji, the Great Pyramids, Popocatepetl and Palenque in Mexico, and the Ganges River.)

We arrived at the Radisson Poco Diablo Resort expecting incense-filled tents and teepees, drumming circles, and Grateful Dead CDs. But there was none of that. Instead we found an older crowd of people wearing clothes from Target and T.J. Maxx. They would fit right in at the outlet mall near our home.

Going to this event felt a bit like going to a casino. At most casinos we have visited (quite a few, recently!), attendees aren't happy-go-lucky vacationers enjoying themselves. Rather, many seem worried and desperate to win. You can't help but wonder if the person standing next to you is having the worst day of his or her life. The psychic fair had a similar feeling of desperation. Many people seemed to possess the vain hope that a psychic would help them recover from some major personal disaster. As scientists with the professional and personal perspective that every single service and product offered at the fair amounted to so much horse manure, we found it really depressing.

The wares were astonishing. You could buy "quantum accelerated" flashlights, pendants, and laser pointers to shield you from the negative frequencies of cell phones, laptop computers, and radio waves. You could acquire a silicone bracelet to bring you balance, health, and power by "aligning the protons of your body" (which, if true, would turn you into a magnet, though the vendors were not aware of this fact). Creams and ointments to pull the negative energy from wounds and cancers were on sale. A man named Elvis had a Polaroid camera in a box with a rapidly changing color-wheeled lamp inside. It produced a picture of Steve surrounded by mystical colored blobs, which Elvis explained were Steve's guardian angels, spirits, and energies. Elvis smiled and said, "Thirty-three dollars, please."

The rest of the vendors were psychics or astrologers who did readings for a fee of anywhere from $15 per fifteen minutes to $35 per half hour. Some used tarot or other types of cards, some grabbed your hands and went into an immediate trance, and some gave either Western or Asian style massages to rid the body of negative frequencies. Did you know that fifth-dimensional quantum healing is extremely effective in third eye and DNA activation?

To avoid the possibility of a hot reading, we did not reveal our last names (to ensure that the psychics couldn't simply perform an Internet search and dig up facts about us), nor did we give detailed information about ourselves.* Our cover story was that we had come for advice on how to raise our kids in harmony with nature in this crazy technology-ridden world. We also wanted to find out more about an object that Susana owns. We have a small collection of miniature toy soldiers, some of which are quite old and valuable. The latest addition is an aluminum toy depicting a British soldier from the 1760s on the march with his musket on his shoulder. Susana had found it as a prize inside a Kinder chocolate egg purchased from a sweets shop sometime in the 1980s.

But we didn't tell our four psychics any of that. Instead, Susana told them that she had found the soldier lodged between two planks of wood in her old rented apartment in Boston and that she felt a special connection to the toy. She also told them that she was thinking of going back to school (Susana has a PhD in medicine and surgery, no further schooling required) and asked for advice on what to do.

Each of the psychics had a slightly different method of reading Susana's mind. Some plied cards; others hummed as they held her hands from across the table. They looked at her intently, held the little soldier, and concentrated to accomplish the "psychometric" readings of the toy's history and significance.

*While some psychics might argue that they cannot perform accurately when clients lie to them, we make a counterargument: Shouldn't a psychic be able to see the truth nevertheless? If they can divine how many children you had four thousand years ago in a previous life, seeing through an inaccurate statement should be child's play.

SPOILER ALERT! THE FOLLOWING SECTION DESCRIBES MAGIC SECRETS AND THEIR BRAIN MECHANISMS!

HOW TO DO A COLD READING

Teller says that a cold reading involves teasing out information from a client with questions phrased as statements. "I sense you've got an issue or problem that's concerning you." Of course they do. Otherwise why would they be there? Everybody worries about health, money, love, and death. So if you say, "I sense some problem with your health" and they don't respond, you continue without skipping a beat: "I don't mean your physical health. It's your emotional . . . or financial health." And so forth. Every statement is made with rising inflection, grammatically a statement, but inviting completion as a question. You miss a lot of the time, but people forget the misses and remember the hits.

Flatter your subject shamelessly. Remember, the psychic succeeds by telling you what you want to believe. Ham it up. Don't blurt out "You like ice cream." Look deeply into the crystal ball, the guy's palm, tea leaves, tarot cards, food stains on his shirt—whatever—and slowly show an expression of insight and discovery: "Your rising moon in the Milky Way tells me you like ice cream." The bigger the ham, the more he'll swear by your powers. "That's amazing! I *love* ice cream."

Base questions on the client's stage of life. According to the mentalist Derren Brown, people in their twenties tend to be quite self-involved, wondering what their real self consists of. Older people may be more worried about illness and death. Make empty truisms—"You are sometimes introverted"—sound substantial. Leave everything wide open. For example, you might say, "You are very creative, but it may not be that you specifically, say, *paint*, it may be that your creativity shows itself in more subtle ways." If the person paints, bingo, you're a mind reader. If not, you are flattering the person's inner creativity.

Always ask: Who's Michael? Or Linda, or a similarly common name that the victim will likely match or suggest a variation, like Mike or Mitch, Lynn or Lynette, etc. Never go with "Who's Bathsheba?" Unless, of course, you've nailed everything 100 percent and want to aim for a grand slam finish. Remember, as a psychic, you're not constrained by either time or the truth. If there's no one in the present with that name, you ask if they ever knew someone in the past, and if that fails, shift to the future with a worried expression, saying, "Be careful when you meet someone named Bathsheba, I sense difficulties, possibly a betrayal . . ."

END OF SPOILER ALERT

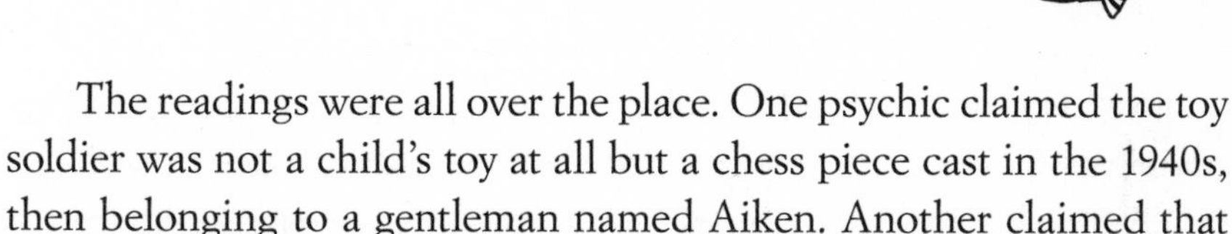

The readings were all over the place. One psychic claimed the toy soldier was not a child's toy at all but a chess piece cast in the 1940s, then belonging to a gentleman named Aiken. Another claimed that the toy was a German soldier (it's a British soldier) and that the connection Susana feels is due to the fact that Susana lived in the same place in Germany that the soldier was from, when she was a scullery maid in a past life. Another said that the toy depicted Susana from a previous life. "It's you when you were Caesar, no wait, one of Caesar's generals. Who was Caesar again? Was he some kind of king?"

As for Susana's immediate future, two psychics said her guardian spirits wanted her to go back to school, while the others noticed that we came to the fair with our two young children and said the spirits wanted Susana to stay home with them. Given that Susana spent fourteen years on her education and has never considered giving up her career, both prognostications were ludicrous—but they could have been reasonable bets for many new mothers.

We concluded that if magicians are artists of attention and awareness, psychics are poseurs of false wizardry. The ones we met in Sedona showed little insight or sophistication. Their method was to ply and probe clients to determine their desires and then, for a price, sell them the promise of those desires. The industry is thriving because people are desperate to confirm that everything is going to be all right, that their decisions have been good ones and will continue to be good ones, and that they will be reunited with their loved ones on the "other side."

How can you defend yourself against psychics, frauds, high-pressure salespeople, priests, politicians, and anybody else who uses cold-reading techniques to get your money? We are not claiming that all people who choose these walks of life are consciously fraudulent. Many believe in their methods and genuinely feel they are helping lost souls. If people walk away from a psychic feeling better about themselves, then no harm done. But some psychics are scammers who use cold-reading techniques to lie to you and take your money. As in all commercial ventures, buyer beware.

Mentalists and psychics often tell you exactly what you want to hear. The psychics who read Susana's "future" changed their story based on her body language and facial expressions. When she smiled and nodded the "clairvoyants" were encouraged to expound on a particular topic, but when she raised or knitted her eyebrows, they would revise the preceding statement. "I see success in your future" one of the psychics said. Susana frowned and tried her best puzzled expression. "Not professionally" the "psychic" immediately corrected, "I mean you will have successful, meaningful personal relationships." Susana smiled and relaxed her shoulders. The self-proclaimed visionary also relaxed visibly.

Some salespeople use similar methods to "read your mind." Next time you go buy an expensive item and suspect the seller is being less than truthful, try changing your story along the way: for instance, tell the salesperson that you are interested in the safety features of a specific car rather than in its design, then change your mind after a while and say that you are really interested in design more than in safety. If the car's best selling points change according to your stated needs, then the salesperson is not honest about the product but is telling you what you want to hear.

One final question: Why, if cold reading is so lame, do people buy it? What makes your brain vulnerable to all the flattery and linguistic legerdemain? People fall for it because, in fact, cold reading is a fundamental component of all human social interactions. Normal polite discourse demands that we seek to determine the needs of our interlocutors in any conversation. We aim to be sensitive, to be charming, to be good listeners. It's how we treat each other civilly. Psychics capitalize and expand on this natural tendency.

PSYCHIC BLUNDERS

Houdini was an early debunker of frauds and charlatans in magic and science, and he took part in a committee overseen by *Scientific American* magazine to scientifically investigate so-called psychics. Houdini's skeptical fervor came from his own previous desperate attempts to contact his dead mother. He tried multiple venues to speak to her; all failed. In one of them, the medium (the wife of Sir Arthur Conan Doyle) famously channeled Houdini's mother for him. She told Harry how much she loved him and how proud she was of him. Too bad Harry's real name was Ehrich and his mother only ever spoke to him in German. Disillusioned and embittered by the experience, Harry Houdini took it upon himself to expose mediums and psychics as mere tricksters.

An iconic vision of the menacing magician involves placing a hapless person from the audience into a hypnotic trance. Svengali. You are getting *sleeee*py. A scam, right?

Not so fast. According to our colleagues who study the brains of people who are prone to trancelike states, hypnosis is not necessarily hocus-pocus. The age-old practice profoundly alters neural circuits involved in perception and decision making, changing what people see, hear, feel, and believe to be true. Recent experiments led people who were hypnotized to "see" colors where there were none. Others lost the ability to make simple decisions. Some people looked at common English words and thought they were gibberish.

The experiments were led by Amir Raz, a cognitive neuroscientist at McGill University in Montreal, who is an amateur magician. We've never met him, but we like him already. Raz wanted to do something really impressive that other neuroscientists could not ignore. So he hypnotized people and gave them the Stroop test. In this classic paradigm, you are shown words in block letters that are colored red, blue, green, or yellow. But here's the rub. Sometimes the word "red" is colored green. Or the word "yellow" is shown in blue. You have to press a button stating the correct color. Reading is so deeply engrained in our brains that it will take you a little bit longer to override the automatic reading of a word like "red" and press a button that says "green."*

Sixteen people, half of them highly hypnotizable and half of them resistant, came into Raz's lab. (The purpose of the study, they were told, was to investigate the effects of suggestion on cognitive performance.) After each person underwent a hypnotic induction, Raz gave them these instructions:

> Very soon you will be playing a computer game inside a brain scanner. Every time you hear my voice over the intercom, you will immediately realize that meaningless symbols are going to appear in the middle of the screen. They will feel like characters in a foreign language that you do not know, and you will not attempt to attribute any meaning to them.
>
> This gibberish will be printed in one of four ink colors: red, blue, green, or yellow. Although you will only attend to color, you will see all the scrambled signs crisply. Your job is to quickly and accurately depress the key that corresponds to the color shown. You can play this game effortlessly. As soon as the scanning noise stops, you will relax back to your regular reading self.

Raz then ended the hypnosis session, leaving each person with what is called a *posthypnotic suggestion*—an instruction to carry out an action while not hypnotized. Days later, they entered the brain scanner.

In highly hypnotizables, when the instruction came over the intercom, the Stroop effect was obliterated, Raz said. They saw English words as gibberish and named colors instantly.

But those who were resistant to hypnosis could not override the conflict, he said. The Stroop effect prevailed, rendering them significantly slower in naming the colors.

When the brain scans of the two groups were compared, a distinct pattern appeared. In the hypnotizables, Raz found, the visual

*See http://sleightsofmind.com/media/strooptest.

area of the brain that usually decodes written words did not become active. And a region in the front of the brain that usually detects conflict was similarly dampened. Top-down processes overrode circuits devoted to reading and detecting conflict. Most of the time people see what they expect to see and believe what they already believe—unless hypnosis trips up their brain circuitry. Most of the time, bottom-up information matches top-down expectation, but hypnosis creates a mismatch. You imagine something different, so it is different.

The top-down nature of human cognition goes far to explain not only hypnosis but also the extraordinary powers of placebos (a sugar pill will make you feel better), nocebos (a witch doctor can make you ill), talk therapy, meditation, and magical stagecraft. We are not saying that hypnosis can cure your cancer, but these effects all demonstrate that suggestion can physically alter brain function.

Magicians use suggestibility, hypnosis, and the illusion of choice to control the path of our behavior during a performance. We come away mystified as to how they could have known what we would do in a given situation, when in fact they were controlling our minds the whole time.

Decades of research suggest that about 10 to 15 percent of adults are hypnotizable. Up to age twelve, before top-down circuits mature, 80 to 85 percent of children are highly hypnotizable. One in five adults is flat-out resistant to hypnosis. The rest are in between, prone to occasional hypnotic states such as losing all sense of time and surroundings while driving on a monotonous highway or watching a spectacular sunset. No one knows what makes one person more or less hypnotizable, although certain subtypes of a gene called COMT may confer susceptibility.

But those who are susceptible can be identified with the help of standard questionnaires and interview techniques. Many are complicit in that they believe hypnotism is effective. They expect it to work, so it does.

Brain scans show that the control mechanisms for deciding what to do in the face of conflict become uncoupled when people are hypnotized. They are then open to suggestion. Thinking that a medicine will relieve pain is enough to prompt the brain to release its own natural painkillers. People who expect pain not to be as bad as it actually is experience a reduction in pain equivalent to that achieved by a shot of morphine. Hyperactive children who are given a "dose extender" in full knowledge that it is an inactive pill can reduce their regular medication by half with no ill effects. Hypnosis and placebos are effective anesthetics. They are used for treating anxiety, tension, depression, phobias, addictions, asthma, allergy, high blood pressure, and many other medical conditions.

In all these instances, top-down processes override bottom-up information. People think that sights, sounds, and touch from the outside world constitute reality. But the brain constructs what it perceives based on past experience.

Hypnosis provides a window into exploring the human condition. We all color reality based on our experiences, expectations, suggestions, and beliefs. The fact that these are shaped in large part by culture, family upbringing, advertising, peer pressure, and spiritual inclination is fodder for many fascinating future studies.

Paul Zak, a neuroscientist, amateur magician, and director of the Center for Neuroeconomics Studies at Claremont Graduate University in Claremont, California, likes to tell a story about himself when he was a teenager. Crazy about cars, Zak took a job at a gas station on the outskirts of Santa Barbara, California. "You see a lot of interesting things working the night shift in a sketchy neighborhood," he says. "I constantly saw people making bad decisions: drunk drivers, gang members, unhappy cops, and con men. In fact, I was the victim of a classic con called the pigeon drop."

Zak recalls that he met a man coming out of the restroom with a pearl necklace. "Found it on the bathroom floor," the guy said. "Geez, looks nice. I wonder who lost it?" Just then the phone rang and another man asked if anyone had found a pearl necklace. He had just bought it for his wedding anniversary. He offered a $200 reward for the necklace's return. Zak, feeling happy to help, told the man that a customer had just found it. "Okay," the man said, "I'll be there in thirty minutes." Zak gave the gas station's address and the man gave his phone number.

But all was not well. The man who found the necklace said he was late for a job interview and could not wait for the other fellow to arrive. What to do? "Hmm. Why don't I give you the necklace and we split the reward?" Zak felt his greed-o-meter go off in his head, suppressing all rational thought. "Yeah," said Zak, "you give me the necklace to hold and I'll give you a hundred bucks." The deal was made. Zak, who earned minimum wage, didn't have a hundred dollars, so he took the money out of the cash drawer—just as a loan, of course.

The rest is predictable. The man with the lost necklace never showed up. He did not answer phone calls. Finally, Zak called the police, who told him that the necklace was a two-dollar fake and that his calls had gone to a pay phone nearby. Deflated, Zak confessed to his boss and repaid the money out of his next paycheck.

Zak, today a leading authority on the neurobiology of trust, is interested in finding out why cons such as the pigeon drop work. He also wonders why people suspend their disbelief in the presence of magicians.

The answer may lie in *oxytocin*, the hormone released during childbirth, breast feeding, social recognition, and cooperation. Zak and his colleagues have carried out numerous studies showing that oxytocin makes acts of cooperation feel really, really good. When you feel trusted, your brain releases oxytocin, and that causes you to reciprocate the trust. If you inhale oxytocin in a laboratory experiment, your generosity to strangers skyrockets.

Zak asserts that con men and magicians are equally adept at causing your brain to squirt oxytocin to make you trust them. But in this seduction they use different techniques and have different ends in mind.

The key to a con, says Zak, is not that you trust the con man, *but that he shows he trusts you.* Con men ply their trade by appearing fragile or needing help, by seeming vulnerable. Because of oxytocin and its effect on other parts of the brain, you feel good when you help others. "I need your help" is a potent stimulus for action. As for the pigeon con, the first hook was Zak's desire to help the poor guy get this nice gift to his undoubtedly sweet wife. The second hook was the man who wanted to give the necklace back but who was late for his interview. If only Zak could help him get that job. Zak's oxytocin system was in high gear, urging him to reciprocate the trust he had been shown and to help these people. Only then did greed kick in. "Hey," thought Zak, "I can help both men, make a wife happy, and walk away with a hundred bucks—what a deal!" Yes, suspend all suspicion and give up the cash. Cons often work better when an accomplice poses as an innocent bystander who "just wants to help," says Zak. We are social creatures, after all, and we often do what others think we should do.

BOXING FRAUD—BEYOND A REASONABLE DOUBT

What if you received an e-mail from an analyst who said he had a system for predicting the winners of certain upcoming boxing matches with 100 percent accuracy. He can predict the outcome of only a few fights, but he knows exactly which fights are possible to predict and in those cases the prediction can be made several days ahead of time based on the characteristics of the fighters and other secret factors. He doesn't expect you to believe him and he's not asking you for anything. He'll prove the system to you by sending you the predictions ahead of time. You needn't reply and you can do anything you want with the prediction, including ignore it or make bets on it. For example, in two days there will be a fight: Boxer A will beat Boxer B.

You don't reply to the e-mail, but out of curiosity you check the outcome of the fight online and find that, indeed, Boxer A won.

For the next three weeks you get a new e-mail accurately predicting the outcome of a fight that week. The odds of having guessed the outcome of all four of these fights by chance is one in sixteen. Pretty good!

The following week another e-mail comes but it's labeled with high importance. The analyst has made another prediction, but what is exciting this time is that the published odds on the fight are 10:1 in favor of the guy the analyst says will lose. An additional payout of 2:1 is offered from his bookie for large bets made on the underdog (whom the analyst predicts will win). He needs to make the largest bet possible to maximize the winnings. A $5,000 contribution from you will return $100,000.

That's a lot of cashola, and his betting record looks solid. You follow the instructions in the e-mail and wire the money before scanning the real estate section of the newspaper in anticipation of your winnings.

A few days later, your guy wins! But you never hear from the analyst again. What gives?

There was no system, Poindexter. You got taken. Here's how.

The analyst collects a hundred or so e-mail addresses through Google searches and sends out e-mails like the first one you got above. But there's one small difference. In exactly half of the e-mails, he predicts that the winner will be Boxer B. The next week the analyst sends out only fifty e-mails to the winning recipients of last week (the Boxer A group). Half of these e-mails predict that Boxer C will win this week and half predict Boxer D will win. The next week, only twenty-five e-mails are sent out to the previous week's winners, Boxer D, and so it goes. Finally, you're a member of a select group of six people who get the final prediction and the request for money. Two of you actually send the dough. One of you won and one of you lost, but either way the "analyst" keeps the $10,000.

Our legal system is based on the idea that criminals can be put away if they are proven guilty beyond a reasonable doubt. To a magician or con artist, the concept of reasonable doubt is dubious. Rather, they know that people will accept evidence as ironclad when they fail to perceive that they are being duped. Scams like this show how easy it is to string people along based on their flawed estimation of probabilities.

Bernie Madoff, the king of cons who pulled off the largest Ponzi scheme in history, used private golf clubs and other exclusive establishments to lure investors. He cultivated the illusion that only very special people could invest with him, people he trusted and who in turn could trust him. He played hard to get: "I don't need your money. Investments are risky. I don't know if you want to be in my inner circle." Madoff, in the eyes of his victims, was one of the good guys who championed the interests of the small investor. Meanwhile, he was milking their oxytocin circuits all the way to the bank.

Zak has good advice for how to avoid a con. Oxytocin's effects, he says, are modulated by your large prefrontal cortex that houses the "executive" regions of your brain. Oxytocin is all emotion, while your prefrontal cortex is deliberative. If you know how easily your oxytocin system can be turned on by charlatans, you should, with mindfulness, be less vulnerable to people who might want to take advantage of you. But don't be *too* vigilant, he warns. Oxytocin causes us to empathize with others, and that is the key to building social relationships.

Magicians also elicit oxytocin in the brains of their audiences, but to different ends. They want you to trust them, so they, too, pretend to be vulnerable. Remember Randi's book test? The poor old coot. He was bumbling and lost. He could not read that woman's mind, no matter how hard he tried. Everyone in the theater was oozing oxytocin.

Thus magicians' banter is often about the need for help, says Zak. "I'm not sure this is going to work" or "This is technically impossible" or "I am at great risk." They induce oxytocin release by sucking you into the illusion, and in turn you trust them to lead you out safely. They often touch volunteers called up on stage, put their arms around them, and give them small gifts. Magicians tend to be extremely friendly and, like Mac King, disarmingly innocent. With a magician, you know you're being scammed, says Zak, but you let it happen anyway because it feels so oxytocin good.

11

THE MAGIC CASTLE

Susana looks out into the blackness from a small stage in a tiny pub called the Hat and Hare. This is it, our big night, June 7, 2010, the culmination of our yearlong effort to learn to perform magic tricks. We are here at the Magic Castle, a funky mansion with many pubs nestled in the Hollywood Hills, to try to win entry into the prestigious Academy of Magical Arts as performing magicians—only we are billing ourselves as the world's first neuromagicians. Can we bring it off? Can we convince the panel of nine professional magicians sitting in the dark before us—including Shoot Ogawa, the most famous Asian magician in the world, and Goldfinger, aka Jack Vaughn, the Society of American Magicians Hall of Famer and perhaps the most prominent African American magician in history—that we deserve to be members of their inner circle?*

Cross Hogwarts School of Witchcraft and Wizardry with an English pub and Disney's Haunted Mansion and you'll get the Magic

*The other judges were Goldfinger's partner Dove and Scott Smith, Tim Vient, Allan Rosson, Bill Koppany, Amos Levkovitch, and Mike Elkin.

Castle. The building is the Area 51 of magic and bills itself as the most exclusive club of magicians in the world. This is the sanctuary where many of the world's greatest magicians let down their goatees, hang out, and relax. Once a month, they invite a few wannabe magicians to audition. We are trying out in a group of six people, which is larger than usual. You can't get an audition without a current member sponsoring you, and even then only about half the candidates pass on the first go. Many more are encouraged to try out again after another one to three months of practice. The Castle sometimes provides a mentor to give weekly lessons until the candidate is up to snuff. Those who pass muster are eligible for a Gold Pin membership, which provides access to the extensive library of magical arts, lectures, and shows, plus the right to vote on academy matters.

Over the past year we had been practicing an act that we developed with the help of Magic Tony, our close friend and tutor. In recent months, as our date with destiny approached, we met in Starbucks, IHOP and other breakfast joints, wine bars, and even a large empty classroom in the psychology building at Arizona State University, where Tony is a graduate student. Tony taught us classic tricks—using cards, ropes, bits of paper, Jell-O, and gimmicks—and helped us dress them in modern garb. On stage, we wear white lab coats, with our name and the title "Neuromagician" stitched on the left breast pocket.

In our act, we demonstrate that we can make an exact replica of a person's brain using a special Polaroid camera and a pan originally designed to hold live doves. We provide false explanations of how the technology works using rope tricks and magicians' gimmicks. We read minds. And then, in the end, we perform surgery on the brain, which is made of Jell-O, to extract a playing card that a volunteer has been "holding in his mind" all through the act. Our patter is mostly nonsense delivered with an air of authority and, we hope, humor.

Susana clears her throat and begins, "Hello, ladies and gentlemen, and thank you for coming to tonight's Wonder Show. As you may know, Wonder Shows were one of the ways preindustrial scientists and inventors disseminated their discoveries to the public. In the nineteenth century, photographs were prohibitively expensive, and literacy, for that matter, was not yet ubiquitous. So scientists went on the road to show the wonders of the age and the discoveries that were changing the world."

WHAT WOMEN MAGICIANS?

Susana addressed the audience as "ladies and gentlemen" but there are very few women, at least in the United States and Europe, who make their living performing magic. We have asked many magicians why this is so. The answers we've received are more amusing than illuminating: Women can't lie. Women don't get tricks. Women can't do math. Women can't command respect. Girls don't receive magic sets as birthday presents.

The lack of women in magic is self-perpetuating. Teller points out that fifty years ago there were hardly any women in comedy. Now nearly half of all comedians are women. So the larger issue at play may be the lack of cultural tradition and role models for aspiring women magicians. In Asia, for instance, female magicians are much more common. At the 2009 Magic Olympics in Beijing, Max Maven told us over tea that, historically, Asian women often performed religious rituals involving magic, and that geishas incorporated magic into their elaborate entertainment routines.

Taking his turn at center stage, Steve nonchalantly shuffles a deck of cards. "Wonder Shows are all but gone, now replaced with high-quality publications and TV documentaries," he says. "Which is all well and good. But there is something missing on the page and on the screen that can only be fully experienced with live experiments on innocent vict—uh, that is, I mean . . . real people. Tonight, we will revive the Wonder Show form of scientific discourse. We will show you the wonders of our modern age, with a special emphasis on brain science."

Steve takes a step forward and gazes into the audience. "Let's get started by asking for a volunteer."

Eight jurors point simultaneously to the only other person in the room: the ninth member of the committee, Scotto (otherwise known as Scott Smith, a professional magician with a day job as a quality assurance engineer at the University of Southern California's Marshall School of Business). He's been our primary handler for the audition process, the one who scheduled our tryout and sent us the performance guidelines: *No fire. Have fifteen minutes of performance ready. If you perform as a duo, make sure that each person does enough magic to be evaluated individually.*

The key thing he *hasn't* told us is what the judges are looking for. We assume they want to see skill with sleights of hand, patter, humor, originality, and timing. Only later do we learn the three main requirements that they are judging us on. We must be good enough never to embarrass the Magic Castle. We must not reveal magic secrets through poor performance. And our timing must indicate that we understand when the magic happens for the audience—that we aren't just going through the motions.

MAGIC TRICK CATEGORIES

All magic tricks follow certain central themes:

- Appearance: You produce something from nothing—a rabbit from a hat, a coin from thin air, a dove from a pan.
- Vanishing: You make something disappear—the rabbit, the coin, the dove, the Statue of Liberty, whatever.
- Transposition: You cause something to move from one place to another—as when Tamariz transports cards from a table into the jacket pocket of somebody he's never approached.
- Restoration: You destroy an object, then bring it back to its original state—as when a magician rips up your hundred-dollar bill and then hands it back to you intact.
- Transformation: An object changes form, such as when a coin turns

into a different coin or three different lengths of rope are transformed into three equal lengths.

- Telekinesis (levitation or animation of an object): You defy gravity by making something rise into the air—such as the classic woman with the hoop run around her. Another example is Teller making a red ball hover and follow him around onstage. Or you make a spoon bend with your thoughts alone.
- Extraordinary mental or physical feats or extrasensory abilities: You catch a bullet with your teeth or you can tell what a person will choose. Johnny Thompson's precognition trick from chapter 7 is a good example.

Poor Scotto. He is apparently destined to suffer any abuse we may issue during our performance.

Steve approaches Scotto, saying in a soft, crooning, magician-style voice, "Am I correct that we've never met before tonight and that you are acting of your own free will as my assistant?"

Scotto replies that he's only spoken to Steve through e-mail correspondence as part of the audition process and that Steve has never asked him to serve as a stooge for the act about to unfold.

"Thank you. Then I'll ask you to first choose a card as I riffle through them with my thumb. You can tell me to stop anywhere you like."

Steve cuts the cards and extends his right hand in front of Scotto, running his thumb down the corner of the deck. The click of each card is clearly audible in the minuscule bar. Even the red velvet curtains that cover the walls can't absorb the loud snaps.

About halfway through the deck, Scotto says "Stop." Steve removes the cards above the stopping point and allows Scotto to take the chosen card, which is now on top of the half-deck.

SPOILER ALERT! THE FOLLOWING SECTION DESCRIBES MAGIC SECRETS AND THEIR BRAIN MECHANISMS!

Of course it is a force. We are setting up a complex trick in which we will magically transport a card into the middle of a brain made out of Jell-O. But first we need Scotto to pick a card identical to the one we embedded last night into the fake Jell-O brain. It is the jack of diamonds.

To force the card onto Scotto, Steve loads the jack of diamonds as the top card of the deck, then shuffles the cards without actually moving the jack. When this false shuffle is complete, Steve cuts the cards into his left hand, which puts the jack of diamonds in the middle of the deck, but he sticks his left pinky just above it so that he knows exactly where the card is. From the front of the deck, the cards look flat, but from the back there is a clear gap caused by the "pinky break." A master wouldn't have had to actually stick his finger into the deck. The pinky would simply hold open a small gap. But despite months of practice, it's clear to Steve (and probably everyone in the room) that he's no master.

With the pinky break in place, Steve runs his left thumb down the front corner of deck ("the riffle") and waits for Scotto to say "Stop." But no matter where Scotto chooses to stop, Steve will lift the cards from the back of the deck at the pinky break, ensuring that Scotto's "choice" is the jack of diamonds. Steve's misdirection involves looking into Scotto's eyes as he lifts the cards, so as to keep Scotto's attention away from the sleight of hand.

Learning tricks like these, we've been surprised to discover, is just as much about what you do with your eyes and body as it is about what you do with your hands. The trickiest part for us has been to learn to do things without attending to them—or, more precisely, while attending to something else. Pulling off these simple sleights requires about as much dexterity as you need when learning how to shuffle a deck of cards for the first time. But to learn to pay attention to irrelevant things while specifically not attending to the secret methods—all the while not looking guilty? Very difficult.

If we learned one thing during our magic training, it is that the route to success is practice, practice, practice, and more practice. This is true of every motor skill you acquire throughout your life—learning to walk, kick a soccer ball, play the piano, hit a tennis ball, block a punch in tae kwon do, ski down a black diamond slope, or put a pinky break in a deck of cards. But now we aren't just directing a ball to a specific point at a specific time, we are also using our own spotlight of attention to misdirect.

Human motor skills are countless and often amazing. People born without arms can dress themselves and write letters—with their toes. Contact jugglers, such as David Bowie's character in the movie *Labyrinth*, can manipulate glass balls with their hands and arms to create the illusion that the balls are floating in midair.* Acrobats can do handstands on top of galloping horses. But we acquire all our motor skills in the same way.

You have in your brain swaths of tissue, called the *motor cortex*, that map all the movements you are able to make. Your primary motor map sends commands from your brain down to your spine and out to all your various muscles. When this map is activated, your body can move. You have other motor maps involved in planning and imagining movements, but for now let's look at how a familiar skill develops.

Let's say you are learning to play the piano. When you are a novice, the region of your brain that maps your fingers—yes, you have finger maps—grows in an exuberance of new connections, seeking and strengthening any connection patterns that maximize your performance. If you give up practicing, your finger maps will stop adapting and shrink back to their original size. But if you keep practicing, you will reach a new phase of long-term structural change in your maps. Many of the novel neural connections you made early on aren't needed anymore. A consolidation occurs: the skill becomes better integrated into your maps' basic circuitry, and the whole process becomes more efficient and automatic.

There is another level to all this, and that's true expertise, or virtuosity. If you practice a complex motor skill day in and day out for years on end, always striving for perfection, your motor maps again

*See http://sleightsofmind.com/media/contactjuggler.

increase in size. Professional pianists (and magicians!) unquestionably possess enlarged hand and finger maps. Their maps are larger than average because they are crammed full of finely honed neural wiring that gives them exquisite (and hard-earned) control of timing, force, and targeting of all ten fingers. Violinists also have enlarged hand maps—but only one. The map that controls their string-fingering hand is like the pianists'. But their bow hands, while deft and coordinated, do not become beefed up beyond normal.

Here is one more interesting fact about expertise. As you gradually master a complex skill, the "motor programs" it requires gradually migrate down from higher to lower areas in your motor circuitry. Imagine a guy who signs up for samba dance classes. Like all novices, he is terrible at first. During his first several lessons, he is processing his dance-related movement combinations up in his higher motor regions, such as the supplementary motor area. This area is important for engaging in any complex and unfamiliar motor task. The dance moves are at first very complex for him. He needs to pay attention to them constantly, and even so he often loses track.

He sticks with it, though, and after a couple of months he is getting a lot smoother. He is using his supplementary motor area much less for his dancing these days. Many of the motor command sequences he is using now have been transferred downward in the cortical hierarchy, to reside mainly in his premotor cortex. He's become a competent dancer. He's not Fred Astaire, but he needs to pay less attention to the basics now. He makes far fewer mistakes. He can improvise longer and longer sequences.

Finally, if he practices often for many months stretching into years, eventually his premotor cortex delegates a lot of its dance-related sequences to the primary motor cortex. Now he can be called a great samba dancer. Dance has mingled intimately with the motor primitives in his fundamental motor map. The dance has become part of his being.*

Susana experienced the gradual acquisition of expertise when she practiced the martial art tae kwon do through high school and

*Another structure, the *cerebellum*, located at the base of the brain, is involved in the coordination, precision, and accurate timing of skilled movements.

college. She has a brown belt and was once the junior tae kwon do champion of Galicia, the region of Spain where she was raised. She found that in the sparring ring, novice martial artists baldly telegraph their intentions through eye movements and body language. The same is typically true of new magicians, who need to think about their tricks as they perform them, and therefore perform them badly.

Accomplished magicians don't need to pay attention to their moves during a trick because the movements come as second nature, as naturally as walking or talking, leaving them free to attend somewhere else. Juan Tamariz jokingly asserts that each spectator is a "telepath." He says that if the magician thinks, even for a brief instant, "Here's where I do the trick," the audience will be able to tell. Thus magicians must be able to perform their routines by rote, without needing to engage any conscious processes. If this is accomplished, the audience won't be able to isolate the critical instant or location of the secret method behind the trick. We all do this in real life to some extent. If you have something to hide from your business partner, spouse, or a law enforcement agent, you will do best not to think about it while in their presence, lest your voice, gaze, or posture give you away.

THE FRENCH DROP OR DECEPTIVE BIOLOGICAL MOTION

Arturo de Ascanio, the father of Spanish card magic, once said that sleight of hand must be so good that attentional misdirection is not needed, and that the misdirection must be so perfect that sleight of hand is superfluous.

We've talked a lot so far about how magicians misdirect your attention. But what about sleight of hand? How does a magician learn to perform flawless sleights, and are any parts of the maneuver more important than others?

Sleight of hand involves making your hand movements ambiguous so that it looks like you are doing one thing when in fact you are doing another. For example, the "French Drop" is a classic sleight in which a coin is apparently removed from one hand by the other and then moved to another position in space before revealing that the coin has disappeared. The moves take a lot of practice to perfect, but nobody has examined scientifically the critical aspects of the maneuvers, until now.

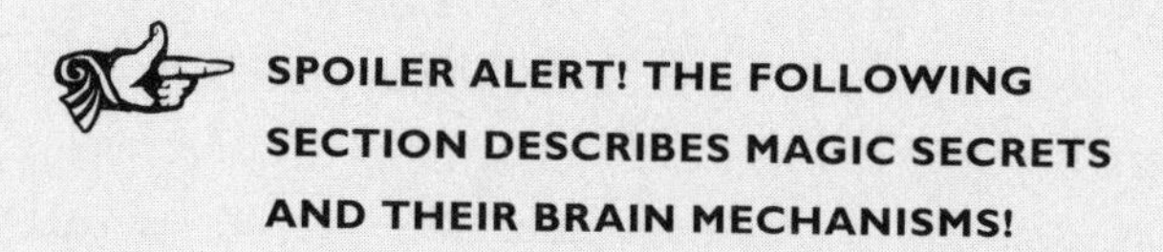

In this famous vanish, the magician holds a coin in one hand and moves his other hand as if to grab it. But instead of taking the coin, he drops it into the palm of the hand holding it and uses his grabbing hand to provide cover. When he moves his grabbing hand away (which you are sure holds the coin), you soon see that it is empty. In fact, the coin is hidden in the palm of his holding hand in a way that makes the hand seem empty.

END OF SPOILER ALERT

Michael Natter and Flip Phillips, researchers in the Department of Psychology and Neuroscience at Skidmore College, recently studied the French Drop by showing videos of both novice and expert magicians performing the trick. They split the movements into three phases: the Approach phase, in which the grabbing hand is approaching the holding hand; the Mid-Capture phase, in which the grabbing hand appears to capture the coin; and the Retreat phase, in which the grabbing hand appears to move away with the coin.

Which phase is most important to the successful sleight of hand? The scientists asked naive observers to watch the videos of the individual phases of the sleight and guess which hand held the coin at the end of the video. They discovered that the Approach phase was not critical to the sleight. The subjects were unable to guess by watching Approach videos from either novice or expert magicians. The Mid-Capture phase, however, was critical. Here, subjects usually guessed the final position of the coin when novice magicians performed the trick but not when experts performed it. The same was true for the Retreat phase, though the effect was not as big as in the Mid-Capture phase.

These results suggest that skilled magicians are more proficient than amateurs in making ambiguous hand movements during the Mid-Capture portion of the trick. They are so good that the parts in your brain that perceive biological motion cannot tell the difference between a real grab and a fake grab.

Steve is talking to Scotto. "Now, there's no need to keep that card to yourself. Show it around as we set up our first technological demonstration, our first installment of the Wonder Show. Whatever you do, Scotto, it is critical that you *keep your card in mind* throughout the show. Some of the technology depends on it."

Susana hands Steve a Polaroid camera.

Steve says, "To ensure you don't forget, we'll take a picture of you and your card. Okay, hold your card right up next to your face, facing me. Good. Think about the card and say cheese." Steve presses the shutter and the camera spits out a Polaroid image.

Steve turns to Scotto again and says, "This Polaroid camera has been specially modified to image your two brain hemispheres. We call it the 'Hemi-roid.' We can use the image to create an exact replica of your brain. Please remain seated while your Hemi-roid develops."

The picture shows Scotto with his card held to his face. But in silhouette over his forehead a line drawing of a brain has appeared.

To make this happen, we placed a transparency of a brain over the film box, between the lens and the film, within the Polaroid camera. Thus all of the images taken with the camera have a big black line drawing of a brain superimposed. The trick here is to know how to line up the brain image with the head of the subject. Like everything else, it takes a bit of practice.

It must be said, we are a bit wooden in our acting skills. It's one thing to get up in front of a group of peers and talk about research.

We've done this enough that public speaking is second nature. The problem we have with our act is the script. When we speak about science, we can make up the specific wording as we go. But with the magic act, there are specific lines that must be said in a specific order and with specific inflections and emotions. Acting is a critical skill for a magician. Robert-Houdin once said, "A magician is an actor who pretends to have real powers."

Susana approaches Scotto while Steve returns to the stage. "May I have your card?" Scotto hands it over. "The memory of your card is now engraved in your brain, and we also have a picture of the card—and your brain—so we can simply dispose of the actual physical card," says Susana as she rips the card into little bits. "But just for further reminder, I'll give you a little receipt to hold on to." Susana returns a card fragment to Scotto.

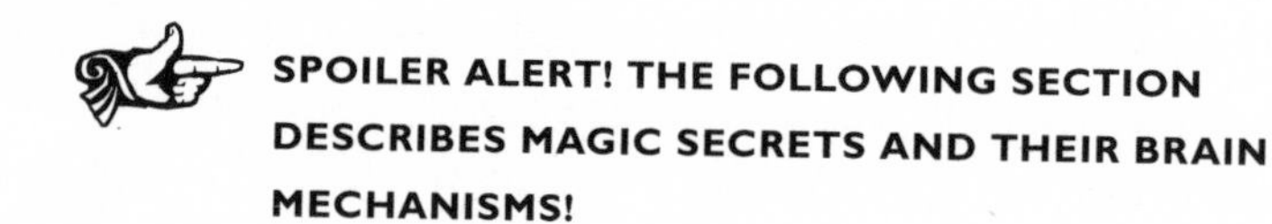

Why rip up Scotto's card now? Because while Susana is ripping up his card, she carries out a classic sleight in magic—the *switchout*. She is secretly holding the fragment from the duplicate card—the one inside the Jell-O brain—between her index and middle fingers. Once Scotto's card is completely ripped, Susana then hands Scotto the fragment from the brain card, as if it came from the newly torn jack. Later, when we remove the jack from the brain, Scotto will find that the fragment he is holding impossibly and exactly matches the missing corner. Pure teleportation!

It took Susana several multihour lessons with Magic Tony, and two or three destroyed decks of cards, to perfect the sleight. She does it brilliantly in the audition, raising her gaze to look Scotto in the eye and misdirecting his attention at the critical time of the switch. Scotto will tell her later that he knew she must be performing a switchout when she ripped his card, but nevertheless he couldn't detect it when it happened.

After she rips the card, Susana returns to a little table near center stage, which supports a crystal goblet. "Remember to keep your card in mind," says Susana, as she deposits the bits of Scotto's card into the glass and covers it with a drape.

Steve puffs himself up and announces, "Ladies and gentlemen, Susana will now introduce the highlighted technology of our show. It's the Digital Optical Volumizing Electronic Positron-Accessing Neuroprinter—or, for short, the DOVEPAN."

SPOILER ALERT! THE FOLLOWING SECTION DESCRIBES MAGIC SECRETS AND THEIR BRAIN MECHANISMS!

This is a joke designed for an audience of magicians. A *dovepan* is a gimmick made of two nested pans with a large covering on top—roomy enough to hold live birds, birthday cakes, you name it. You can buy them in every magic shop. The magician displays the bottom pan, which is empty. He covers it and then waves his magic wand. The top pan drops down into the bottom pan automatically by virtue of a spring-loaded mechanism that is activated when the top and the bottom pan meet. He then removes the cover. Voilà, a dove flies out. Or a rabbit hops out. Or—you guessed it—a Jell-O brain appears. It looks like magic.

END OF SPOILER ALERT

Our dovepan rests on a small table, covered by a surgical drape, to the right of the stage. We embellished it with a huge handle and various electronic devices bulging from its top. Mad science: check.

Steve says, "The DOVEPAN will now analyze Scotto's hemorrhoid—er, Hemi-roid—and use it to create an exact replica of his brain."

Steve approaches Scotto and says, "May I grab your Hemi-roid?" Finally we hear a few snickers from the serious (not easy to please, not easy to fool) crowd. We both think this is a good sign.

Steve holds up the photo to the jury and hands it to one member to pass around, saying, "Notice that this Hemi-roid is a true and factual representation of Scotto's brain." He retrieves the photo and mounts it on the dovepan. Susana says, "And now the dovepan will use the Hemi-roid to make an exact replica of Scotto's brain!"

Susana rubs her hands, mad-scientist style. "We'll need to add raw materials to build a brain," she says. "A brain needs lots of fat." Steve grabs an ice cream scoop, scrapes a large dollop of Crisco from a bucket, and flings it into the bottom portion of the dovepan. The scoop strikes the pan's edge, ringing it like a bell.

Then Steve says "We need protein" and hands a large carton of body-building protein powder to Susana. She peels off the lid and shakes "protein" into the pan. Next Steve picks up a full sugar dispenser that he recently stole from a truck stop. Susana pours it all into the pan and declares, "Sugar!"

"And now, most important, salt," says Steve. "Salt is critical because its ions—sodium and chloride—allow neurons to communicate over long distances." Steve unscrews the top of a saltshaker and, with great exaggeration, pours a stream of salt into his left fist.

"The neural signals travel from this end of the neuron"—he moves his right hand along the pathway of the activity from his left hand, up his left arm, and across his chest to Susana's waiting outstretched hand—"all the way over to the postsynaptic neuron, represented by Susana's right hand."

At this point, Steve suddenly begins an incredibly dorky rendition of the classic break-dancing step in which a wavelike motion begins at the end of one arm and flows through the other arm.

"This process is called 'saltatory conduction,' " says Steve, as they hold hands and the wave continues through Susana's body.

SPOILER ALERT! THE FOLLOWING SECTION DESCRIBES MAGIC SECRETS AND THEIR BRAIN MECHANISMS!

When Steve's right hand travels across his body, he is actually retrieving a fake thumb tip—a rubber gimmick that looks just like a real thumb—that was used to sequester the salt inside his left fist. His right hand delivers the fake thumb tip to Susana's waiting gloved left hand. When the undulating duet is done, Steve removes Susana's left glove, which serves to get rid of his thumb tip. Susana is also wearing a fake thumb tip filled with more salt in her right hand, under her glove. At the end of her dance number, she removes her right glove, palms the thumb tip into her right fist, and pours her salt supply into the dovepan.

It appears as if the salt has traveled through two bodies.

END OF SPOILER ALERT

With the salt now in her right hand, Susana says, "And now, the postsynaptic neuron has been activated." She pulls up her lab coat lapels and moonwalks over to the dovepan as Steve plays a Michael Jackson tune on his iPhone. Steve hears a few members of the jury say, "Choreography!" as if crossing one performance element from a list.

The music stops as Susana pours salt into the dovepan. She places the lid on the device, presses the three-second timer button, and says, "Now, we wait."

As we embellished our dovepan before heading off to the Magic Castle, we realized that essentially, in magic, there are no new tricks. Nearly all the illusions you see in modern magic shows were invented in the nineteenth century or earlier by showmen in Europe, Asia, and the Americas. Modern magicians have been updating and elaborating the same basic tricks ever since.

Moreover, magicians have long excelled at engineering. In the second century BC, Heron of Alexandria, a Greek-Egyptian inventor, made temple doors open and close magically during religious ceremonies. The secret mechanism was a predecessor to the steam engine. Magicians also used to be famous for inventing self-operating machines, called *automata*, with purely mechanical moving parts. For example, in 1739, Jacques de Vaucanson invented the digesting duck, which appeared to have the ability to eat kernels of grain, metabolize the grain, and defecate.*

*The food was collected in an inner container and the prestored "feces" were produced from a second container.

"Heron's Temple." Heron of Alexandria invented the automatic opening of doors. The secret mechanism, called aeolipile, consisted of a vessel with two curved pipes connected to it. When the water in the vessel boiled, the steam came out of the tubes, activating a rope mechanism that opened the doors slowly and majestically. (Illustration by Victor Escandell for the Fundación "la Caixa" museum exhibit "Abracadabra, Ilusionismo y Ciencia")

In the mid-nineteenth century, Jean-Eugène Robert-Houdin, who is considered the father of modern magic (and the main inspiration for Ehrich Weiss, better known as Harry Houdini), used his engineering skills as a clockmaker to construct amazing mechanical contraptions that seemed to operate by magic. A device similar to two different famous Robert-Houdin automata called "Orange Trees" is featured in the 2006 movie *The Illusionist*. Robert-Houdin also invented the first electric house security alarm and other Rube Goldberg contraptions such as a three-tiered alarm clock system that set off alarms at different places around the house and at different times while also triggering the release of morning oats to his mare in the barn. Other renowned magicians, such as André-Jacques Garnerin and John

Nevil Maskelyne, made important technological advances by inventing the parachute (Garnerin) and the first ribbonless typewriter and the coin-operated lock for vending machines and, unfortunately, pay toilets (Maskelyne).

We read Robert-Houdin's 1860 autobiography—*Memoirs of Robert-Houdin, Ambassador, Author, and Conjurer, Written by Himself*—to learn more about this period. This guy's life story reads like a rip-roaring Victorian novel. One of his tricks stands out as an example of how devious magicians are and how little has changed over the past century.

While visiting a prominent local sheikh at a remote desert compound, Robert-Houdin demonstrated his bullet trick. Penn & Teller have a killer bullet trick that is based on this earlier version.

In the trick, which he demonstrated to large audiences in Algiers, Robert-Houdin dared a volunteer from the audience to shoot him point-blank. Having prepared his apparatus in advance, he "caught" the bullet in his teeth.

But here, in the desert, Robert-Houdin was taken by surprise. A skeptic challenged him then and there: "I will lay out two pistols. You choose one. We will load it and I will defeat you."

Robert-Houdin had to buy time. "I require a talisman in order to be invulnerable," he replied. "I have left mine at Algiers. Still, I can, by remaining six hours at prayers, do without the talisman and defy your weapon. Tomorrow morning at eight o'clock, I will allow you to fire at me."

The magician spent two hours that night ensuring his invulnerability. He took a bullet mold out of his pistol case. Then he took soft wax from a candle, mixed it with a little lamp black, and made a wax bullet. He hollowed it out so it would not be hard. Next he made a second ball and filled it with blood. Robert-Houdin later explained that an Irishman once taught him how to draw blood from the thumb without causing any pain.

The next morning, Robert-Houdin stood fifteen paces from the sheikh, who held the loaded pistol. The gun went off and the bullet appeared between Robert-Houdin's teeth. Furious, the sheikh lunged for the second gun, but Robert-Houdin reached it first. "You could not injure me," he said, "but you shall now see that my aim is more dangerous than yours. Look at that wall." The Frenchman pulled the trigger and on a newly whitewashed wall there appeared a large splotch of blood.

Robert-Houdin had used sleight of hand to put the wax bullet into the first gun, and it broke into pieces when fired. He held a real bullet in his mouth and—voilà. With equal dexterity, he placed the blood-filled bullet in the second gun before firing it. The sheikh nearly fainted.

THE MECHANICAL TURK

The first magical contraption to become world famous was "the Turk," an automaton that played master-level chess, invented by the Hungarian baron Wolfgang von Kempelen in 1769. Spectators were welcomed to see the calculating machinery inside its box after each show. Stories about the Turk, especially who discovered its secrets, are legion. One account states the real workings of the Turk were revealed in 1827, when two skeptical young boys from Baltimore hid and watched backstage as a man climbed out of a hidden compartment. The local newspaper broke the story that the chess-playing "automaton" was a hoax.

Perhaps we'll never know the full truth, but Robert-Houdin's account of its origin is as plausible as any. He writes that in 1769, a revolt broke out in a half-Russian half-Polish army regiment stationed at Riga, in what is now Latvia. The leader of the rebels was an officer named Worousky, a man of great talent and energy. Troops were sent to suppress the revolt, and in the rout both of Worousky's thighs were shattered by a cannonball. He threw himself into a ditch behind a hedge and at nightfall dragged himself to the adjacent house of a kindly physician named Osloff. After gangrene set in, both of Worousky's legs were amputated.

Not long after, Wolfgang von Kempelen, a celebrated Viennese inventor of mechanical devices, visited Osloff. Together they devised a plan to help Worousky, who had a bounty on his head, to escape. Worousky was a brilliant chess player, which gave von Kempelen the idea for an automaton chess player. In three months, they built the device—an automaton represented as the upper body of a Turk seated behind a box the shape of a chest of drawers. In the middle of the top of the box was a chess board.

Before each game, von Kempelen opened the doors to the chest so people could see various wheels, pulley, cylinders, springs, and so forth. The Turk's robes were raised so the "body" could be inspected.

After closing the doors, von Kempelen wound up one of the wheels with a key. The Turk nodded its head in salutation, placed its hand on one of the chess pieces, raised it, and deposited it on the board. The inventor said the automaton could not speak. It would signify "check" to the king by three nods and to the queen by two.

The legless Worousky was stowed away in the body of the legless Turk. As soon as the robes fell, he would enter the Turk's upper body, passing his arms and hands into the figure and his head into the mask.

According to Robert-Houdin, the magical machine gave Worousky an escape and a livelihood. The Mechanical Turk toured Europe extensively and won nearly all of its chess matches.

"Turk automaton." The operator could hide under the shell of the automaton. (Illustrations by Victor Escandell for the Fundación "la Caixa" museum exhibit "Abracadabra, Ilusionismo y Ciencia")

Throughout the nineteenth century, magicians were at the forefront of technology and invention, but at some point the development of new effects essentially stopped and magicians clung to their (now) old traditions and technologies. Much of the low-hanging fruit had been plucked, and it was easier to continue to do the same old tricks. More recently, a few magicians, such as Jason Latimer, the winner of the world championship of magic (FISM) in 2003, have embraced modern technologies—lasers, holography, fiber optics, electronics, robotics—and used them to make wholly modern magic and live onstage special effects.* The basic effects on the brain are still the same (to the best of our knowledge, they haven't developed truly new categories of magic effects yet), but they make fresh and exciting new variants on old tricks using high technology.

MAGICIANS AND SPIES, UNITE!

In 1952, the CIA asked one of the nation's most respected magicians, John Mulholland, for help. Could the master close-up sleight-of-hand artist teach American spies a trick or two in their escalating cat-and-mouse game with Soviet spies?

The reasoning made sense. Both spies and magicians must elude detection. The CIA's many dirty tricks—poison darts, knockout powders, drugs, poisons, tiny cameras—would be operationally useless unless field officers and agents could manipulate them. If Mulholland could deceive an audience that was studying his every move from a few feet away, it should be possible to use similar tricks for secretly administering a pill or a potion to an unsuspecting target.

Mulholland obliged by writing two illustrated spy manuals. The first describes and illustrates (with delightful drawings) numerous sleights of hand and close-up deceptions for secretly hiding, transporting, and delivering small quantities of liquids, powders, or pills. The second manual

*See http://sleightsofmagic.com/media/jasonlatimer.

describes methods used by magicians and their assistants to secretly pass information.

The George Smileys* of the day embraced the techniques and, to read modern accounts, became adept at misdirection, change blindness, escapology, and creating cognitive illusions. As the Cold War heated up, the CIA's field officers grew ever more inventive under Mulholland's guidance.

By the 1970s, however, attempts to assassinate Fidel Castro with exploding cigars and similar escapades began to embarrass the CIA. In 1973, the agency's director, Richard Helms, ordered all copies of the classified magic manuals to be destroyed. The results of such chicanery were just too unpredictable.

For decades, rumors of the manuals' existence circulated in intelligence circles, until parts of them were unearthed and published in the late 1990s and early 2000s. In 2007, a retired CIA officer, Robert Wallace, discovered a complete set of the lost manuals and published them, with the historian H. Keith Melton, under the title *The Official CIA Manual of Trickery and Deception.*

The book reveals that our spies knew about change blindness. An intelligence officer would always park his car at the curb directly in front of his house. On the day a "drop" was to be left for another agent, the officer would park his car across the street from his house. The agent would notice this and pick up the secrets, but the enemy's surveillance team would not see anything out of the ordinary.

This ploy was successful in Moscow, home to the heart of the KGB's surveillance operation. The American intelligence officer would adopt unvarying patterns of daily movements in and around the city. After a few months of this unchanging travel pattern, the American spy would "disappear" during his "normal" commute for a brief time—enough to accomplish a dead drop or post a letter—before reappearing at his normal destination only minutes behind schedule. The watchers, lulled by the monotony of his routine, were not alarmed.

*Smiley is the main character in John le Carré's novel *The Spy Who Came In from the Cold.*

In magic, a larger action covers a smaller action as long as the larger action itself does not attract suspicions. One CIA officer took his dog out for long walks at night (the large action), which gave him numerous opportunities to secretly mark signal sites and service dead drops (the smaller actions). The surveillance teams became used to the pattern and never got suspicious.

Magicians manage "sight lines" to create illusions. Your vantage point in the audience can be used to trick your visual system, as we saw with Vernon's Depth Illusion in chapter 2. A CIA officer discovered that when he was walking in urban areas, on routes he used frequently, the surveillance team trailing him was always a few steps behind. When he made a right-hand turn on foot, he would be in the clear—"in the gap"—for a few seconds. He used that gap to conduct his clandestine moves, out of sight.

Mulholland also gave lessons on misdirection. In the days when many people smoked cigarettes, he instructed officers to lift a flaming match to light a target's cigarette while using the other hand to drop a pill into the target's drink.

To make a miniature camera "disappear" after taking a secret photo, the spies borrowed a magician's tool called a *holdout*—a simple piece of elastic that retracts an object up a sleeve. They hid toolkits and microfilm in buttons, coins, boot heels, and suppositories.

Houdini inspired many of the spies' techniques, including the Identical Twin Illusion (which they called "identity transfer"), which involves disguising two people to look like the same person. One spy went a step further and dressed up in a giant Saint Bernard dog suit so that when he was "taken to the vet" (actually a safe house) he could pass on documents before returning home in the dog suit. A real 180-pound Saint Bernard also lived there.

When the timer chimes, Susana lifts the lid of our dovepan and reveals that the ingredients have been transformed into a human brain. Well, not a real brain, but as realistic as one made of Jell-O can look.*

*Our Jell-O brain was very realistic indeed. We rushed to our Magic Castle audition through heavy LA traffic with the car's air-conditioning on full blast and the vents wide open to keep the brain in Susana's lap as cool and structurally sound as possible.

JELL-O AND MAGIC

We made a human brain out of Jell-O using a classic Halloween brain recipe.

1. Spray a small amount of cooking spray inside a plastic brain mold.*
2. Place contents of 2 large boxes gelatin mix (peach or watermelon) into a large bowl.
3. Add 2½ cups boiling water. Stir gelatin with a whisk until it is completely dissolved, about 3 minutes.
4. Stir in 1 cup cold water.
5. Add 1 can nonfat evaporated milk and stir for 2 minutes.
6. Add a few drops green food coloring (to make the brain grayish pink); stir.
7. Pour gelatin mixture into plastic brain mold.
8. Set mold in refrigerator overnight.
9. Stick card into brain once it's solid while still in mold. The small entry point cut will be unnoticeable on the bottom of the brain.
10. Add cerebral arteries using sparkly red cake decorating frosting.

*These are available at http://www.shindigz.com/party/Gory-Brain-Mold.cfm.

Steve says, "And here we have it, ladies and gentlemen—an exact replica of Scotto's brain!" Steve removes it from the dovepan and places it on a second small table, visible to all.

"You all must be asking yourselves, how does this incredible DOVEPAN technology work?" says Susana. "Well, it's based on genetic manipulation, leading to rapid neural growth, directed by the model provided by Scotto's Hemi-roid."

At this point, we each carry out a rope trick to illustrate various

---

Susana joked about a possible car crash in which a team of very confused paramedics would arrive at the scene to find both Steve and Susana unconscious (but with intact skulls) and the brains of a missing third person splattered on the asphalt.

aspects of how the DNA is manipulated in the dovepan so as to rapidly grow an exact replica of Scotto's brain. The strands of rope represent strands of DNA, and our scientific explanations are nutty, but we handle the ropes okay.

We are feeling pretty good about the show. A little more than halfway through, we've completed the trickiest sleights in the act. The methods thus far have been standard magic fare, and we are entering the portion of the show with the cool mentalism tricks.

So it comes as a shock when one of the jurors says, "I think I've seen enough."

We are now in a much larger bar upstairs at the Magic Castle, consoling ourselves with expensive Perrier-Jouet champagne. Magic Tony joins us. We tell him that we have just been summarily dismissed from our audition, halfway through. Now we know how those poor talentless saps from *The Gong Show* felt. But we are determined to celebrate, no matter what. We are so embarrassed that we are overtaken by the giggles, like that poor Spanish politician who had admired the firefighters' "equipment."

The conversation inevitably turns to what went wrong. We know we are no Penn & Teller, but we do think we achieved what we set out to do. A few minor rough spots, to be sure, but nothing horrifically bad. Did we fail to earn their trust? Were we an embarrassment to the professionalism of Magic Castle members? Did we flub our tricks?

Disappointingly, we didn't even get to show our coolest tricks! The rest of our act is a humdinger. Here's what we had planned.

We bring two volunteers on stage and have them play our version of a mentalism puzzle called *kirigami*, invented by Max Maven. It involves folding and cutting paper with letters of the alphabet to find four-letter words. The volunteers think they are free to find a variety of words, but we have set up the puzzle to force them to choose only two: "cage" and "head."

We bring out homemade "mind-reading helmets" constructed out of spaghetti strainers adorned with flashing lights and buzzers—they look like Acme bombs purchased by Wile E. Coyote—and each push a secret remote button in our jacket pockets to make the helmets buzz as the volunteers concentrate on their words, which are being "transmitted" through the air to the dovepan.

After three seconds, Susana lifts the cover of the dovepan and what do you see? Why, it's the confluence of the words "head" and "cage": our technology has generated the head of the actor Nicolas Cage! (It's amazing what you can buy on the Internet.)

Finally, Susana uncovers the goblet containing the card bits but finds that the pieces are missing. They have been replaced by brain matter—bits of Jell-O. Susana takes a taste just to be sure. Yep—definitely human brain matter.

The goblet actually consists of two halves separated by a double-sided mirror. One half contains the card shreds and the other the brain matter. Susana spins the goblet under the cover of the drape when she wants the transformation to take place.

We express puzzlement at this unexpected event. If the card has turned into brain matter, then what happened inside the brain?! We tell Scotto that we must perform exploratory surgery on his Jell-O brain to find out.

In the brain, we find Scotto's card with a piece missing, which, of course, exactly matches the jagged edge of the fragment he still has in his hand. Scotto literally kept his card *in mind*, and our devices produced a replica of his brain, memories, thoughts, and all!

If only they had let us finish! We're sure that this finale would have impressed the judges.

Just then Tim, the head of the committee, approaches us in the bar. Yes, there were a few problems with our act—we definitely shouldn't quit our day jobs—but there's nothing to keep us from joining the Magic Castle as Gold Pin members.

In response to our confused expressions, he says that they cut our act short because we showed proficiency and they had four other auditions to do that night.

"Congratulations!" he said, shaking our hands.

"We made it!" we whooped, clinking our glasses.

12

WILL THE MAGIC GO AWAY?

Now, because it is relevant, and witchcraft so apparently accomplished through the art of sleight of hand, I thought it would be worthwhile to explain it. I am sorry to be the one to do this, and regret any effect this may have on those who earn their living performing such tricks for purposes of entertainment only, whose work is not only tolerable but greatly commendable. They do not abuse the name of God in this occupation, nor claim their power comes through him, but always acknowledge what they are doing to be tricks, and in fact through them unlawful and unpious deceivers may be exposed.

—Reginald Scot, *The Discoverie of Witchcraft*, 1584*

When we tell people that we are studying the neuroscience of magic, the same questions invariably come up: Was it difficult

*Modern English text by Neil Alexander.

to get magicians to reveal their secrets? After all we have learned, do we still enjoy magic? By explaining how magicians hack the human brain, do we worry that we'll ruin the mystery for everyone else? Will the magic go away?

We have been fortunate to work with some of the world's greatest magicians who have been generous in sharing their ideas about the essence of magic and, yes, often willing to reveal their secrets. The reason is that great magic is not about secrets. Nor is it all about the tricks or the methods behind the tricks. You can find complete descriptions and explanations on the Internet of just about every magic trick ever invented.*

A great magician makes you experience the impossible by disrupting normal cause-and-effect relationships. Sure, he can use secret methods, but his act will be even more magical if you know the secret and yet the impossibility still occurs. Successful magicians hijack your brain's attentional mechanisms without your knowing it—you believe you've been paying attention the whole time. No matter what trick they are doing, the real trick is in your head, so secrecy is not as important as many believe.

As we have noted before, magicians are masters of live performance who have spent thousands of hours practicing their art. We learned to do a few magic tricks pretty well, but we are not good enough to expect anyone to pay to watch us. Consider an analogy to live music performance: anyone can learn to play a Beatles song on the guitar, but not everyone can be Paul McCartney. Being a great magician involves many things, and knowing the secrets behind certain tricks is only one of them.

Noel Daniel, the editor of Taschen Books and author of *Magic: 1400s–1950s,* writes, "Magic does something really that no other

*Magic secrecy fundamentally ended in the sixteenth century, with the publication of *The Discoverie of Witchcraft.* This book, which reveals that performance magic is achieved by natural means, was intended as an argument against the existence of witches, and as a protest against witch hunts. Today, the magic publishing industry is huge. Amazon.com currently sells 79,119 books on magic (almost seven times more than "romance novels," currently 11,653 on the Amazon site). That's not even counting the instructional DVDs. Many more YouTube videos disclose magic tricks and provide step-by-step instructions on how to perform them.

kind of performing art can do, and that is, it manipulates the here and now—our reality. When we're watching a movie, we don't think that what we're watching is real. We know it's not. We stare in a dark room at a lit screen. But in magic, we're watching someone manipulate a coin, or cards or fire or sawing a woman in half, right on stage, right in front of our very eyes. And this is the power of magic."

Many successful magicians have told us that "exposure," or giving away secrets, will not be a problem for their own business. They have their show, their public, their fans, and they have no qualms about talking shop with scientists. Many of these folks sell magic books and trick sets directly to the public in big-name toy stores and bookstores as well as in the gift shops of their own shows. But they nevertheless exercise caution, because magic is their livelihood. In addition, if a magician is perceived as giving away too much—as lifting the veil of secrecy—he might be shunned by the magic community at large.* It's not worth the risk.

Indeed, this is a contradiction—it's okay to give away secrets and it's not okay to give them away—and we sympathize with magicians caught in the middle. Throughout this book we have used spoiler alerts to warn readers whenever secrets are about to be revealed. We did this to ensure that we adhere to the letter of the ethical guidelines of the magicians' associations to which we belong, which insist that the public must not learn a secret by accident. We are members of the Academy of Magical Arts, the Society of American Magicians, the International Brotherhood of Magicians, and the Magic Circle in England.

The various organizations representing magicians consider exposure a punishable ethical violation and have guidelines to determine if a magician is guilty of malicious exposure. These guidelines appear designed to protect the public from the "ravages" of magical knowledge, as if protecting virgins from carnal knowledge. Or maybe it's more about protecting the bottom line. Modern ethics statements stipulate that secrets should be distributed only in return for payment

*The popular TV series *Magic's Biggest Secrets Finally Revealed* featured an anonymous masked magician because of fear of repercussions.

(selling a book, teaching a lesson), lest the secrets run rampant through society and make magic shows impossible. Ironically, no magic associations that we know of have ethics committees dedicated to protecting the public from false claims of paranormal abilities by magician members.

Sitting outside the Rio Hotel and Casino in Las Vegas late one night after his show with Penn, Teller described the code of ethics as an outdated mind-set. It's as if magic were some sort of medieval guild that needs to guard its secrets, he said, transmitting its esotery only from master to apprentice. In fact, Teller was personally criticized by some of his peers for exposing, in step-by-step photographs, the Miser's Dream trick in a *New York Times* article on our neuromagic collaboration.* Teller doubted he hurt any magician's business.

MAGIC BOYCOTT

David Pogue, the *New York Times* technology writer, recently wrote a story about a curious iPhone app called iForce.† The application presents itself as a drawing program called Doodle v1.2, but it's really a sophisticated trick that uses the iPhone's internal accelerometers to create a mentalism effect based on precognition. After you buy this app for $3, you write a prediction on the iPhone screen, using your finger in a painting app. You lay the phone facedown on the table. You ask your friend to choose a number between one and eight. Or to pull a bill out of his wallet. Or flip a coin three times and remember the sequence of heads and tails.

You then ask your friend to tell you the number, show the bill, or reveal the coin toss sequence. You turn the iPhone back over, and—will wonders never cease—that is exactly what you wrote on the screen. It might be seven, $20, or tails, tails, heads. Your prediction was correct.

†Pogue's Posts, "The Magic Behind Rating Apps," *New York Times*, May 27, 2010.

*The photos are similar to those in chapter 10, page 192.

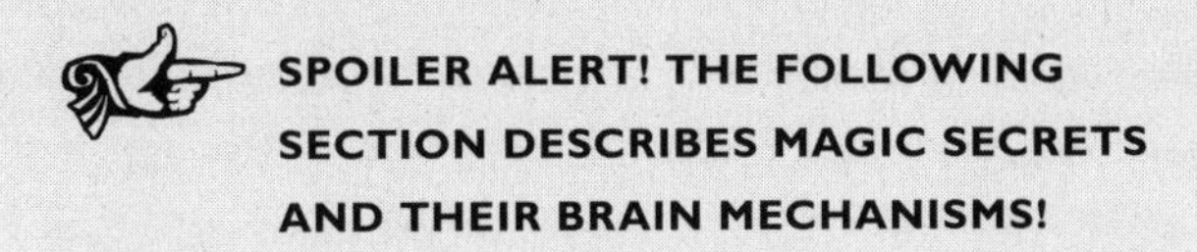

SPOILER ALERT! THE FOLLOWING SECTION DESCRIBES MAGIC SECRETS AND THEIR BRAIN MECHANISMS!

The app works because when you appear to be making the prediction, you are really running two fingers side by side down the face of the iPhone, and that opens a secret screen in the program. You select what type of trick it is (numbers 1–8, type of bill, coin toss, and so on) and then set the phone facedown as if to hide your prediction. When you flip the phone over to reveal the prediction—note that there are only eight possible answers to the questions—you can flip it to the left or right, over the top or the bottom, fast or slow. In other words, you have eight possible ways of flipping the phone faceup, depending on your friend's answer.

The phone interprets the way it is being flipped, and the iForce app draws the correct response on the screen.

END OF SPOILER ALERT

Oddly, the app's creator, Grigor Rostami, noticed that not long after the program was launched his ratings on iTunes went from an average of five stars (the best) down to an average of three stars (middling). When Pogue checked the bad ratings, he found that people were saying the nicest possible things about the app.

"This app is amazing!" (one star)

"Awesome app! One of the best and funniest apps ever. Great job!" (one star)

"Wow—the best $3 I have ever spent! Keep up the low ratings!" (one star)

So why would such a great app get super evaluations but low star ratings?

Rostami started reading magic forums and discovered that magicians were intentionally giving his app a one-star rating (the worst) to keep it secret from everyone else in the app store. The magicians conspired to reduce his sales by sabotaging the star ratings. This would reduce exposure and keep the trick viable (for them) as long as possible. Verbal evaluations don't affect the rating that the app has in the listings, so they were honest in their written praise.

Rostami, who is a magician, told Pogue that once the one-star ratings began coming in his sales dropped substantially.

This kind of screw-the-competition mentality is unfortunately pervasive among some magicians who regard exposure as magic's highest sin. Yet the professional magician organizations issue no ethical guidelines against this kind of behavior. It's an indicator that the overwhelming concern about exposure may fundamentally be an ultimately self-destructive drive to maximize one's own take while reducing the success of others—as if magic were a zero-sum game.

Of course, some tricks are more resilient to exposure than others. For example, the dovepan we used in our act at the Magic Castle is a classic gimmick. Once you know how it works (see chapter 11), you lose your sense of wonder. It's cheapened, no longer intriguing.

But Teller describes a trick that became more intriguing the more he knew. This is the famous Cups and Balls, a sleight of hand that was performed by Roman conjurers as far back as two thousand years ago. The trick has many variations, but the most common one uses three balls and three cups. The magician makes the balls pass through the bottom of cups, jump from cup to cup, disappear from a cup and turn up elsewhere, turn into other objects, and so on. The cups are usually opaque and the balls brightly colored.

Teller recalls that one day he was sitting in a diner in the Midwest with Penn, fiddling with an empty water glass and wadded-up paper napkins for balls. He turned the glass upside down and put a ball on top, then tilted the glass so that the ball fell into his other hand. The falling ball was so compelling that it even drew his own attention away from his other hand, which was deftly and automatically loading a second ball under the glass. He was so well practiced that he no longer needed to consciously control his hands. In fact, Teller found that the sleight happened so quickly he himself did not realize he had loaded the transparent cup. The great magician had misdirected himself!

THE ILLUSION OF EXPOSURE

Apollo Robbins is onstage with Susana, discussing magic and the brain at the Chicago Cultural Center. He's stuffing a large silk handkerchief into his fist. With one eyebrow raised à la Dr. Spock, he's showing the audience how their angles, meaning their sight lines, are critical to successful magic. The audience feels as though it is learning secret magic techniques. It's exposure as entertainment.

When the silk is fully crammed, Apollo opens his fist and, voilà!, the handkerchief has been transformed into an egg. He then pulls the silk from his hip pocket, as if the silk had magically transported itself there from his hand.

"It's an easy trick," he explains. "All you need is a fake egg and two identical silk handkerchiefs." He turns the egg around and reveals that it has a hole into which he stuffed the silk. The crowd laughs as he slowly pulls the stuffing back out through the hole.

"Here's why the angles are important. First, the setup," Apollo says. He refolds one of the silks and puts it in his hip pocket, along with the fake egg. He puts the other silk in his jacket breast pocket. He is now set to repeat the trick.

"Step one, I palm the egg," he says as he secretly extracts the egg from his pocket. "But you can't see that from where you're sitting." His hand is in an ice-cream cone eating position, egg tucked neatly within. "Then I take the silk from my breast pocket and stuff it into the egg like this." Again he stuffs the silk into his fist. "Make sure nobody is behind you or has an angle that allows them to see the egg in your hand."

"Here's the egg just like before," he says, opening his fist. The silk is gone, as expected. "But if you look closely, you can see there's another way to keep people from seeing the hole." He now turns the egg, revealing the small opening. Then, to everyone's amazement, he peels the hole off the egg, showing that it was not an actual hole but a sticker. Yet the handkerchief is gone! Apollo now removes the silk from his hip jacket pocket and flicks the sticker away. To prove that the egg is real, he grabs a glass from the table, and cracks the egg into it.

Magicians call the silk-to-egg trick a sucker trick—the magician does a trick and then apparently exposes its secret method, only to immediately show that the explanation was bogus. It's similar to apparent repetition except now the audience thinks it knows how the trick is done. Sucker tricks are based on apparent exposure, rather than the actual exposure done by the Masked Magician.

Magician Whit Haydn says that if the exposure is better than the routine, expose away. One reason sucker tricks are popular could be that exposure deepens the audience's appreciation for the art of magic and for the skill and cleverness of the performer—even if the exposure itself is an illusion.

Teller further realized that all of this took place despite the fact that he should have been able to see the secret ball as it was loaded under the cup. Its image was on his retina, but he nevertheless missed it because his attention was so enthralled with the falling ball. He surmised that if it worked for him with a transparent cup, it would work with an audience. The transparency of the cups would make the trick all the more magical to the audience. And that is how Penn & Teller came up with the idea for a cups and balls routine using transparency. They claim that their version of the trick violates four rules of magic: don't tell the audience how the trick is done, don't perform the same trick twice, don't show the audience the secret preparation, and never perform cups and balls with clear plastic cups. The exposure is what makes this trick a superstar.

At the magic symposium in Las Vegas, Teller told the scientists that "the core of a successful trick is an interesting and beautiful idea that taps into something that you would like to have happen. One of the things I do in our live show is to squeeze handfuls of water and they turn into cascades of money. That's an interesting and beautiful idea. The deception is really secondary. The idea is first, because the idea needs to capture your imagination."

There's another reason, aside from fairness to their colleagues, that magicians should be generous in revealing their methods:

magic can help increase the rate of discovery in brain science. The discovery of inattentional blindness and change blindness in recent decades (detailed earlier in chapter 5) has greatly advanced the cognitive sciences. Magicians evidently knew implicitly about these phenomena for centuries, judging from the design of their tricks, and so scientists have been inadvertently reinventing the wheel. By studying magic, scientists could have made these advances earlier. We propose that the study of magic is now poised to help to derive new principles to optimize attentional resources in people with cognitive decline, as well as to create heuristics to improve education in our schools.

In another example of magic helping science, David Copperfield, through his foundation, generously designed and funded a program called Project Magic. Teams of magicians and occupational therapists work together to teach sleight of hand to physically challenged people to help with their rehabilitation and improve their self-esteem.

If magic can promote scientific discovery and clinical practice, then magicians might be morally obligated to make their secrets available for use. We're not saying they should give up their knowledge for free. Like all experts, they deserve acknowledgment and remuneration for their creativity and invention. Perhaps they could think of it as enlightened self-interest.

A few years ago, neither of us had ever been to a magic show, nor had we given this ancient art a nanosecond of attention. But now that we understand how it works, we are unabashed groupies. The more we learn about magic, the more interested we become as consumers. We go to magic shows whenever we get the chance, because we love being fooled, even though we have read explanations of many of the tricks. Experiencing a master magician fling our attention around like a fly fisherman's lure, forcing us to strike at the morsel and then reeling us in, is unlike any other cognitive experience we've had outside of the science we do in our labs. It's as though somebody took all of the cool things we study every day and suddenly made them beautiful and dramatic. We like some shows so much that we go to them over and over and never come away disappointed. We have traveled the world meeting magicians, learning from them, collaborating with them, and racking our brains for ways to explain what they do and how they do it. We've taken magic lessons and bought thousands of dollars' worth of magic paraphernalia.

Returning from the front lines, we can say that having expertise in magic makes it more appealing, not less. To understand why, you need to know a little more about *mirror neurons.* Recall that these are the brain cells that become active when you carry out an action and when you observe another person carry out that same action. When you wave good-bye, mirror neurons in your premotor cortex fire away. When you watch someone else wave good-bye, those same neurons fire, but you don't move your body. In other words, mirror neurons link action and perception.

Your mirror neuron system gets more active the more expert you are at an observed skill. When pianists listen to someone else's piano performance, the finger areas in their primary and premotor cortex increase above their baseline activity. Their mirror neuron systems automatically run the performer's keystrokes in emulation. The same thing does not happen in the brains of nonmusicians. While they can certainly appreciate the music deeply, their experience is inevitably shallower than the pianist's in at least one way, because they are not experiencing what it is like to actually produce it.

The same goes for athletics: the better your own skills, the more deeply you understand the skilled performances you witness. For example, when classical ballet dancers and experts at an Afro-Brazilian art form that combines martial arts and dance called capoeira watched video clips of each kind of dance, the dancers' brains showed distinct patterns. Both disciplines require exact limb positions, choreographed movements, extreme muscle strength, and years of practice. You would think that their mirror neuron activity would be equivalent, yet when ballet dancers observed capoeira movements, their mirror neuron activity was weaker compared to when they watched other ballet dancers—and vice versa. The actions you mirror most vividly are the ones you know best.

We are willing to bet that the same holds true for magicians. If Teller watches Mac King perform a fake coin toss, his mirror neurons are going to have robust responses. If an ordinary muggle watches

Mac do the same trick, she will be entertained, but we suspect her mirror neurons will not respond as strongly.

Now imagine that everybody in the world could perform one trick and perform it well. Would magic suffer from this vast increase in exposure? Would ticket sales to shows fall? On the contrary, the more you learned, the more interesting magic would become, because you, and your brain's motor control pathway, would empathize with the activity more deeply.

We think the enduring mystery about magic is how the brain constructs—and falls for—illusions. In this regard, we hold a minority view among our fellow visual scientists. To the generation that preceded us, illusions were considered errors of perception. The late Richard Gregory, the British psychologist who is widely known as one of the most prolific perception scientists in the world, liked to say that illusions are where the visual system got it wrong.

We disagree. Illusions are not exceptions and they are not necessarily mistakes. They are integral to perception and represent fundamental aspects of your visual and cognitive processing. They are adaptive shortcuts that your brain makes to speed up such processing, or reduce the amount of processing necessary to provide you with the information you need to survive and to thrive, even if the information isn't technically accurate.

Try this for yourself: look at this page indoors, and then take it outside and look at it under direct sunlight. It's remarkable in that it's unremarkable. The page looks exactly the same—black letters on a white background. But how can that be? Depending on the nature of your indoor lighting, there is about one million to twenty million times more light* under direct sunlight than indoor light. Outside, there are millions of times more photons reflecting off the black letters than there were off the white paper inside, so why don't the black letters, when outside, look brighter than white?

Furthermore, the colors of the photons (the distributions of wavelengths) are probably different inside and outside, too. Your visual

*As measured per unit area. We call this *photon density*.

system can see color and brightness only as a function of the numbers of photons and their wavelengths that fall upon your retinas. Thus the page cannot possibly be "white" both inside and outside.

If the photons inside and outside are so different (and we assure you that they are), why does the page look the same in both environments? The answer is that your visual system massages the visual data with two processes called *brightness constancy* and *color constancy*, so that the page looks the same to you under very different lighting conditions. But this is an illusion, which means the physical reality doesn't match your perception. In reality, the book has a different physical appearance* in each environment, even though you see it as the same.

Visual illusions help you survive in a visually complex world when you exit from the cave. They help you recognize ripe versus unripe fruit in the tree or by firelight. Similarly, cognitive illusions help keep you alive. You make assumptions, confabulate memories, and attend to only one thing at a time, because it's an efficient way to navigate the world and to find the resources you need. It's more efficient than the alternative, which is to try to process everything you encounter. Accuracy is usually not needed and it's difficult to achieve. You'd need a much bigger head to hold a brain large enough to be always accurate, and humans already have enough of a problem with childbirth because of the size of our noggins.

Magicians have tapped in to the power of cognitive illusions more effectively than scientists have, though less systematically. The magician's goal is to misdirect you and create a sense of wonder (though some con artists use these same tricks to steal). Our goal is to take magic into the neuroscience laboratory and to use it for evil— No, no, we want to use it to increase the rate of discovery about our cognitive processes. We believe that magical methods will prove invaluable in determining the circuits in the brain that process cognition, as well as in revealing important new perspectives on how the brain functions.

And it could work the other way around, too. We've been planning a collaboration with Mac King, who does such a fantastic fake

*The scientific term is *reflectance.*

coin toss. He's so quick you can't catch him doing it. Mac has shown us how he does it, and it looks almost identical to a real coin toss. He can toss the coin (or fake it) for many repetitions before we're able to tell a real toss from a fake toss.

The goal of our project will be to determine if known principles of visual processing might enhance the perception of a magic trick. For example, can Mac intentionally adjust where people look and increase the feeling of magic? Should he adjust the speed at which he tosses the coin to be optimized to visual processing? And if so, does such adjustment actually help create the illusion? The precise answers to these questions can be obtained only through direct scientific experimentation. By answering them, we will determine if the perception of magic is tied directly to the way we optimally perceive stimuli with our eyes.

In order to determine whether magicians have discovered new perspectives on the brain that scientists have missed, we intend to test their intuitions in our labs. For example, as we described in chapter 5, Apollo Robbins intuits that, in some circumstances, a curved motion is more effective than a straight motion for misdirection, whereas straight motion is more effective than curved motion in other situations. One underlying neuroscientific hypothesis is that curved versus straight motion results in different types of eye movements and that those eye movements have different effects on attention. If this is correct, Apollo's insight may reveal an important new perspective on the relationship between cognition and the oculomotor system.

Will all this science make the magic go away? We believe that the wonder and awe of perceiving magic will no more disappear than did the beauty of the sunrise after Copernicus discovered that the earth is a sphere rotating around the sun. Both revelations—that we are hurtling around the sun and that magic works because our brains are inherently limited—are simultaneously deeply humbling and awe-inspiring. Increased humility deepens the mystery rather than dispels it.

A few years ago, Steve stood on the summit of Haleakala, a sacred mountain rising ten thousand feet above sea level on Maui, to watch the sunrise. His father once ran an observatory on the same spot, shooting lasers at mirrors left on the moon to measure how long it takes for light to return to earth. The round-trip time changed from one measurement to the next because Maui moved as the earth's tectonic plates shifted.*

Before Steve's father died, he asked to have his ashes spread on Haleakala, and so it was that on this particular morning Steve found himself standing there with his dad in mind, watching the sun's rays burn holes through the clouds from 93 million miles away.

Imagine you are standing next to Steve. You are at the highest elevation for that particular longitude of the earth. That means that at the moment of sunrise, you are the fastest-moving people on earth with respect to the sun. You are hurtling toward it at over one thousand miles an hour,† which is more than twice the escape velocity needed to leave earth orbit. If our planet suddenly stopped spinning during that sunrise, and your speed became a thousand miles an hour relative to the earth as well as to the sun, you would "see" nothing but that sunrise until you burned up in the heliosphere on your way to the center of our solar system—approximately eleven years later. Now imagine all this science as you enjoy the incredible raw beauty of the moment, and try not to drop your jaw.

A mundane explanation for that same sunrise is that you are standing at an arbitrary position on a not-very-special planet that happens to rotate once every twenty-four hours so that the local star's solar terminator passes over an island in the middle of one of the planet's oceans every twelve hours. So what. To Steve, the extraordinary scientific facts only enriched what was already a very emotional experience.

And the same is true for magic, or anything else, for that matter. The science adds to the experience, makes it deeper, fuller, more satisfying. When you see a great trick and can sense the effect it's having

*Geologists use these ongoing astronomical data to measure the movements of the earth's tectonic plates. The eastern edge of the Hawaiian plate is part of the infamous San Andreas fault in California and the data show that the plate is moving northeasterly, toward North America. Steve's dad's data will thus help define how soon Los Angeles and San Francisco will be crushed down into the earth in the Cascadia subduction zone.

†The angular velocity of our planet at this altitude near the equator.

on the neural circuits that are at the core of your being, it's as breathtaking as a Haleakala sunrise.

The basis of all science is a fundamental love of and curiosity about nature. Magic profoundly manipulates the nature of our conscious experience. As such, it holds the promise of revealing some of the most compelling scientific discoveries imaginable.

You might wonder whether magic, with all its complexities, its emotional, attentional, and cognitive components, might be too complex to use as a tool to discover the fundamental principles of cognitive neuroscience. But we think that anyone who takes this point of view may not realize that such arguments have been made before in biology, psychology, and physics and proven wrong.

For example, the study of the neural basis of consciousness used to be considered an impossible field of inquiry. Now things have changed and dozens of labs, including both of ours, investigate the activity of neurons in relationship to conscious versus unconscious perception. Christof Koch and Francis Crick,* who championed the neurobiological study of consciousness back when it was considered uncouth, made the analogy to the question of life as a scientific topic. It seemed an impossibly complex problem, until James Watson and Crick's discovery of the structure of DNA showed how straightforward it really was.

Just because you can't imagine how something works doesn't mean it's impossible to find out.

Over and over, in the history of science, the same story repeats itself: a supposedly unapproachable subject matter is shortly afterward shown to be very much approachable. Philosophers such as Immanuel Kant stated that the human mind was not susceptible to measurement, and therefore a science of psychology was impossible. Then Gustav Fechner, a German physicist, contracted an eye disorder that made him resign his professorship at the University of Leipzig. After he recovered, he turned his studies to the quantification of mental processes, discovering the exact mathematical relationship between a physical stimulus and its associated subjective perception, thereby inventing the new field of psychophysics, a fundamental cornerstone of psychology.

*Crick, the codiscoverer of the double helix, wrote the first popular book on the subject of consciousness, *The Astonishing Hypothesis*.

And it's the same issue here with developing a science of magic. If humans can build a machine like the Large Hadron Collider to examine the Higgs boson, the ephemeral particle that is the very basis of mass, it should also be possible to discover the brain mechanisms related to magic.

And if we can do that—if we can understand magic fully at the level of the brain's circuits—we will know the neural pathways underlying consciousness itself.

If there is one thing we've learned from becoming magicians, it's that your attention, awareness, intuitions, and assumptions are fair game. Even we, beginner chumps in the field of magic, are skilled enough to lead you down the garden path and eat your lunch behind your back. So what does it say about your brain that you are so easily fooled?

We've given some answers as to why you (and we) are so gullible: our brains create sensory afterimages, our memories are fallible, we make predictions that can be violated, and so on. But as we reflect on the reasons, we are drawn to one that stands above all others in explaining the neurobiology of magic—the spotlight of attention.

Recall that your visual system has a spotlight of attention. It's the region of your visual perception in which you enhance everything that occurs. But the principle holds true for hearing, touch, other sensory systems, and even cognitive functions—for everything your brain does. Your spotlight is directed to a region of your cortex and enhances the activity carried out in that region.

But attention exercises another effect in your brain, too. It not only increases the neural signals at the center of your spotlight, it also suppresses the activity in the surrounding region. In the visual system this can create a so-called center-surround attention focus in your visual field. You see better at the center, while the surrounding items are suppressed.

In your touch system, attention creates a center-surround spotlight on your skin. Apollo Robbins's tap on your shoulder forces you to pay attention to that particular location, while suppressing the more subtle sensations produced by the removal of your wristwatch a few feet away. And in the cognitive areas of your brain, attention creates a center-surround region in whatever type of space is being computed by that region. You may fixate on a given idea and suppress all others that might compete.

Our research shows that the spotlight affects visual processing from the very first stages of the visual pathway, signifying that it is a very important factor in what you see and don't see. We believe it also determines what you hear, feel, and are aware of in a magic show, and indeed in the rest of your waking life.

Our studies further show that the harder you try to attend to something, the more you enhance it and the more you suppress surrounding information. This suppression versus enhancement dynamic gets really interesting when you think about decision making and the role of intuition versus rational thinking.

Malcolm Gladwell in his book *Blink* extols the virtues of decision making based on gut level intuitions. In one example, he tells of a museum that purchased a statue. The institution had experts examine the statue for three months, and they declared it to be authentic. But then the curator showed the statue to an archeologist, who took one look at the piece and advised them, "Try to get your money back." Indeed, the statue turned out to be a fake. The archeologist was able to spot immediately what prolonged scrutiny by committee had failed to detect.

On the other hand, Christopher Chabris and Daniel Simons in their 2010 book, *The Invisible Gorilla*, argue that you should rely on deep rational thinking, not your intuitions, to guide your decisions. For example, some parents choose not to vaccinate their children because of their deeply held intuition that vaccinations lead to autism. Chabris and Simons argue that the apparent link is no more than an illusory correlation. Rational examination reveals that there is no causal relationship between vaccinations and autism.

Who's right? Based on our exploration of neuromagic, we believe that both approaches are correct if you combine their ideas in light of the neurobiology of attention.

In terms of its underlying brain mechanism, an intuition may result from weak neural activity in a given brain circuit. The activity is not strong enough to be accessible to your logical mind and drive your rational decision-making processes.

Brain signals can be weak for a number of reasons. The information coming from your sensory or memory systems is sketchy, as in black art, where the contrast between an object and the background is so low that the object is for all intents and purposes invisible.

Or brain signals may be weak because your attentional mechanisms suppress otherwise strong signals. For example, when Apollo Robbins pulls a quarter out of your breast pocket and moves it elegantly along an arc across your face, you follow it the way a tennis spectator follows the ball. You miss that Apollo simultaneously removes your reading glasses from the same pocket, directly under your nose, even though the image of his stealing hand is falling directly on your retinas.

In this sense, rationality and intuition are two ends of a continuum, with weak (intuitive) signals at one end and strong signals, which can be used to reason with, at the other end. Attention can serve to change the strength of any signal up or down anywhere along this continuum. Thus no decision is purely rational, because even though you see clearly in the center of the spotlight, there is darkness just outside the spotlight. Not only are you influenced by your biases, expectations, and assumptions, but you also actively suppress and ignore critical information. Conversely, the vaguest intuitions and gut feelings usually become accessible to your "rational" mind when you cast your attentional spotlight on them, making them more salient and easier to examine.

The yin and yang of attention affect all your decisions. For example, when we opened our first labs, we jointly hired a technician who cried during her job interview, worried that she would be homesick. We ignored our intuitions that this was not a good sign and relied on the fact that she was, on paper, perfectly well suited and experienced for the position, and that she explicitly told us she wanted the job despite the emotional outburst. The outcome proved unhappy for her and for the rest of our two labs. If only we had analyzed all of the information we had, instead of suppressing the nonrational bits, we might have made a better, more productive decision.

A crucial take-home lesson from this journey through neuromagic is that when you are confronted with the uncertainty of a complex decision with lots of variables, you cannot always anticipate what will turn out to be most important factor, because of the suppressive and enhancing effects of your own attention. To overcome this, you must cast your attentional spotlight over each detail of the decision in turn, even if some initially appear insignificant or ephemeral. Reasoning things through is critical, but so is addressing your intuitions, so that your attentional spotlight can focus on each morsel and bring it forward for analysis. Only then will you be able to see the whole picture.

After our years of living magically, we will never watch a magic trick the same way again. Our appreciation of magic has been deepened, and it's been given gravitas to the *n*th degree by the knowledge that all of magic, every little sleight, is really happening in our minds. We've learned that misdirection and other illusions are important to us humans both on and off the magic stage.

We will now reveal one final secret. In a way, we've misdirected you, the reader, at every step. You may have purchased this book to read about magicians and tricks, sleights and secret methods, but all along you've really been learning the fundamental neuroscience at the center of your being. And that's where all the magic really takes place: inside a three-pound lump of flesh, your own brain.

EPILOGUE

Lessons for Life: Bringing the Magic Home

By tricking us so thoroughly, magicians have taught us to think about neuroscience in new ways. Here are some of the lessons we have learned from them that you can use in your own life. We will post more of these on our Web site http://sleightsofmind.com as we continue to gain insights, so please check back often for updates.

1. Magicians know that multitasking is myth and so they use a "divide and conquer" approach with attention. They split your attention so you cannot concentrate fully on any part of the stage at a given time. When you have a long list of things to do, you may feel tempted to do two or more tasks simultaneously, such as answering e-mails while attending a staff meeting. Chances are you will do neither task well. For your best performance, do one thing at a time.
2. Magicians know that memory is fallible and that the more time that has elapsed between the acquisition and the recovery of that memory, the less accurate it is. Know this about

yourself and keep records of important information and conversations *immediately* after they happen.

3. Even though magicians make mistakes all the time, they set them aside and keep moving forward, and the audience hardly ever notices. You should do the same.
4. Some salespeople and psychics will "read your mind" by telling you exactly what you want to hear. Next time you go buy an expensive item and suspect the seller is taking you down the garden path, try changing your story along the way. For example, tell the salesperson that you are most interested in the contrast and brightness of your next TV. Once a model has been shown to you, inform the salesperson that actually you are most interested in the longevity of the device. If the best selling points of the current model change according to your request, then the salesperson is not being honest about the product, and is telling you what you want to hear.
5. Magicians use humor and empathy to lower your guard. If you sympathize with a magician, you will enjoy yourself more and be less vigilant about catching the secrets behind his magic. When negotiating interpersonal, professional, or business relationships, do as a magician and disarm with charm.
6. Each spectator is a "telepath." If you have something to hide from your business partner or spouse or a law enforcement agent, you will do best not to think about it while in their presence, lest your voice, gaze, or posture give you away.
7. Magicians know that attention enhances one small part of the world, while suppressing everything else. When making a difficult decision such as hiring somebody or taking a job offer, make a list of all the tidbits of information you have, no matter how unimportant they may seem. Then sequentially focus your full attention on each item and consider them each individually and fully. Carefully consider the ramifications of each fact and each feeling or intuition you may have. In turn, your attentional processes will enhance each particular issue, while suppressing all other information. Once you reach the end of the list, you will have a full picture based on both the hardcore facts and your gut feelings. You will be ready to decide.

NOTES

1. The Woman in the Chameleon Dress

2 *neuromagic*: See S. Martinez-Conde and S. L. Macknik (2008), "Magic and the brain," *Scientific American* 299: 72–79.

3 Margaret Livingstone's work: M. S. Livingstone (2000), "Is it warm? Is it real? Or just low spatial frequency?" *Science* 290: 1299.

8 By definition: See the recent special issue of *Scientific American* in which we discuss how our visual perception is dominated by illusions. S. Martinez-Conde and S. L. Macknik (2010), "The Science of Perception Special Issue," *Scientific American Special* 20(1).

10 To get at the neural correlates of PTSD: Read more at http://www.smithsonianmag.com/science-nature/How-Our-Brains-Make-Memories.html.

11 This is where you first detect the different orientations: The discovery of neurons selective to line orientation won the 1981 Nobel Prize in Physiology or Medicine for David Hubel and his partner, Torsten Wiesel. Once orientation selectivity was discovered, the field of visual neuroscience set out to categorize all of the various types of features encoded by the visual system.

13 You also make up a lot of what you see: A receptive field is a region of space that, when acted upon by a particular stimulus, will cause that neuron to respond. It is the part of your retina that each neuron can see. Haldan Keffer Hartline won a Nobel Prize in Physiology or Medicine in 1967 for showing that the retinal neurons that transmit information to the brain respond best to those parts of a visual scene containing the edges of objects. By adding, subtracting, or even multiplying receptive

fields, your brain creates a zoological tree of neurons with individual preferences for various aspects of a visual scene or features of objects.

16 This response causes a ghostly image: S. L. Macknik and M. S. Livingstone (1998), "Neuronal correlates of visibility and invisibility in the primate visual system," *Nature Neuroscience* 1(2): 144–49.

17 Auzinger immediately grasped the implications: Ottokar Fisher, *Illustrated Magic* (New York: Macmillan, 1943).

18 Today a black art act: Two brothers, Joe and Bob Switzer, invented fluorescent paint and Day-Glo paint in the 1930s. Joe wanted to be a magician when he was younger and started fooling around with black light that he and his brother learned to make from *Popular Science* magazine. They sneaked into their father's pharmacy and shone light on different chemicals; some glowed brightly. So they mixed chemicals to develop various kinds of paint that fluoresce under ordinary ultraviolet light. Fluorescent pigments seem brighter than standard pigments because they reflect more visible light than they would if they were not fluorescent.

20 "the sky is filled with stars": All celestial bodies, including galaxies, project dots of light smaller than any photoreceptor in your eye. But then how is it that some stars appear bigger than others? The answer is that some celestial bodies are so bright that the extra light they produce reflects off the back of your retina. This reflection in turn excites many more photoreceptors in a larger circular area. The result is that bright stars seem larger.

2. The Secret of the Bending Spoon

32 Two normal depth perception cues: You may be surprised to learn that the depth perception your brain creates by comparing the images in your two eyes (called *stereopsis*) is an illusion, wholly a construct of your mind. Your left eye and right eye convey slightly different views of the world to your brain. If you close your left and right eyes in rapid succession and look at an object, you will see that the object shifts left to right. With both eyes open, your brain triangulates these two images into a single stereo image, which gives you a sense of depth. This is the principle behind stereo-depth illusions such as in the Magic Eye books.

How stereopsis is actually accomplished in the brain remains one of the deepest mysteries of visual neuroscience. We know a bit, but relatively little compared to what we know about how other processes, such as motion perception, are accomplished. We know that the information from each eye remains segregated at the level of your optic nerves. We also know that visual information from your two eyes converges onto the same neurons in your primary visual cortex. This means that certain neurons in this brain region can respond to stimuli from either eye or both eyes. They are binocular.

But where in the brain does vision, based on both eyes, come together? Where is the depth of each object in the scene computed? Where do the images fuse into one seamless experience? We know these things must happen. Otherwise we would have double vision instead of depth perception. In our own labs, we have found that the processes used to derive stereoscopic perception must arise several levels above the primary visual cortex in the visual hierarchy. Finding the exact location is an area of active research.

Stereopsis contributes to Vernon's trick, too, because your two eyes see your card pushed into the deck from different angles. Your brain triangulates these two different retinal images to compute the depth of the card within the deck. It's an illusion, but stereopsis confirms that the card is mid-deck.

38 Tony took advantage: A. S. Barnhart (in press), "The exploitation of Gestalt principles by magicians," *Perception*.

38 Good continuation is so integral to a plethora of brain mechanisms: Ibid.

39 saws a woman in half: This trick can be accomplished in other ways as well. But in all of them good continuation plays a role in the effect.

40 Charles Gilbert and colleagues: M. K. Kapadia, M. Ito, C. D. Gilbert, and G. Westheimer (1995), "Improvement in visual sensitivity by changes in local context: Parallel studies in human observers and in V1 of alert monkeys," *Neuron* 15: 843–56.

40 A second concept behind the spoon illusion: It has been published as the "Dancing Bar" illusion by Peter Tse and Brown Hsieh at Dartmouth College. The neural basis of this illusion has been shown by Christopher Pack, now at the Montreal Neurological Institute. P. U. Tse, P.-J. Hsieh (2007), "Component and intrinsic motion integrate in 'dancing bar' illusion," *Biological Cybernetics* 96(1): 1–8; C. C. Pack and R. T. Born (2001), "Temporal dynamics of a neural solution to the aperture problem in visual area MT of macaque brain," *Nature* 409: 1040–42.

40 To localize the ends of a line: C. C. Pack, M. S. Livingstone, K. R. Duffy, and R. T. Born (2003), "End-stopping and the aperture problem: Two-dimensional motion signals in macaque V1," *Neuron* 39: 671–80.

3. The Brother Who Faked a Dome

43 For further discussion on how visual art and visual science interact, see S. Martinez-Conde and S. L. Macknik (2010), "Art as Visual Research: Kinetic Illusions in Op Art," *Scientific American Special* 20(1): 48–55.

47 Susana's results showed instead: X. G. Troncoso, S. L. Macknik, and S. Martinez-Conde (2005), "Novel visual illusions related to Vasarely's 'nested squares' show that corner salience varies with corner angle," *Perception* 34: 409–20; X. G. Troncoso, P. U. Tse, S. L. Macknik, G. P. Caplovitz, P.-J. Hsieh, A. A. Schlegel, J. Otero-Millan, and S. Martinez-Conde (2007), "BOLD activation varies parametrically with corner angle throughout human retinotopic cortex," *Perception* 36: 808–20; X. G. Troncoso, S. L. Macknik, and S. Martinez-Conde (2009), "Corner salience varies linearly with corner angle during flicker-augmented contrast: A general principle of corner perception based on Vasarely's artworks," *Spatial Vision* 22: 211–24.

49 In 2006 we designed an experiment: X. G. Troncoso, S. L. Macknik J. Otero-Millan, and S. Martinez-Conde (2008), "Microsaccades drive illusory motion in the Enigma illusion," *Proceedings of the National Academy of Sciences of the United States of America* [hereafter *PNAS*] 105: 16033–38.

49 Her expression is often: M. S. Livingstone (2000), "Is it warm? Is it real? Or just low spatial frequency?" *Science* 290: 1299.

51 The Leaning Tower illusion: F. A. A. Kingdom, A. Yoonessi, and E. Gheorghiu (2007), "The Leaning Tower illusion: A new illusion of perspective," *Perception* 36(3): 475–77.

52 The only difference between these two faces: R. Russell (2009), "A sex difference in facial pigmentation and its exaggeration by cosmetics," *Perception* 38: 1211–19.

53 Some stationary patterns: A. Kitaoka, *Trick Eyes: Magical Illusions That Will Activate the Brain* (New York: Sterling Publishing, 2005).

54 We called the new illusion: S. L. Macknik and M. S. Livingstone (1998), "Neuronal correlates of visibility and invisibility in the primate visual system," *Nature Neuroscience* 1(2): 144–49; S. L. Macknik and M. M. Haglund (1999), "Optical images of visible and invisible percepts in the primary visual cortex of primates," *PNAS* 96: 15208–10; S. L. Macknik, S. Martinez-Conde, and M. M. Haglund (2000), "The role of spatiotemporal edges in visibility and visual masking," *PNAS* 97: 7556–60; S. L. Macknik and S. Martinez-Conde (2004), "Dichoptic visual masking reveals that early binocular neurons exhibit weak interocular suppression: Implications for binocular vision and visual awareness," *Journal of Cognitive Neuroscience* 16: 1049–59; P. U. Tse, S. Martinez-Conde, A. A. Schlegel, and S. L. Macknik (2005), "Visibility, visual awareness, and visual masking of simple unattended targets are confined to areas in the occipital cortex beyond human V1/V2," *PNAS* 102: 17178–83; S. L. Macknik (2006), "Visual masking approaches to visual awareness," *Progress in Brain Research* 155: 177–215; S. L. Macknik and S. Martinez-Conde (2007), "The role of feedback in visual masking and visual processing," *Advances in Cognitive Psychology* 3: 125–52; S. L. Macknik and S. Martinez-Conde, "The Role of Feedback in Visual Attention and Awareness," in M. S. Gazzaniga, ed., *The Cognitive Neurosciences* (Cambridge, Mass.: MIT Press, 2009), pp. 1165–79.

4. Welcome to the Show

60 One critical clue: T. Moore and M. Fallah (2004). "Microstimulation of the frontal eye field and its effects on covert spatial attention," *Journal of Neurophysiology* 91: 152–62; Z. M. Hafed and R. J. Krauzlis (2010), "Microsaccadic suppression of visual bursts in the primate superior colliculus," *Journal of Neuroscience* 30(28): 9542–47; N. L. Port and R. H. Wurtz (2009), "Target selection and saccade generation in monkey superior colliculus," *Experimental Brain Research* 192(3): 465–77; J. W. Bisley and M. E. Goldberg (2010), "Attention, intention, and priority in the parietal lobe," *Annual Review of Neuroscience* 33: 1–21.

64 Other times you can shift your attention around: Study by Keisuke Fukada and Edward K. Vogel, "Human variation in overriding attentional capture," *Journal of Neuroscience*, July 8, 2009.

64 Research shows that: G. F. Woodman and S. J. Luck (2007), "Do the contents of visual working memory automatically influence attentional selection during visual search?" *Journal of Experimental Psychology: Human Perception and Performance* 33(2): 363–77.

64 "retinotopic" space: R. Desimone and J. Duncan (1995), "Neural mechanisms of selective visual attention," *Annual Review of Neuroscience* 18: 193–222.

64 Jose-Manuel Alonso: Our work with Jose-Manuel Alonso also showed that a specific kind of neuron is enhanced during attention in the center of the spotlight, while a different kind of neuron is inhibited during attention in the surrounding regions. The neurons with enhanced firing in the center of the attentional spotlight are known to inhibit other neurons, whereas the neurons with suppressed firing in the surrounding regions are critical to determining the direction of moving objects. These results suggest that the role of top-down attention in the very earliest stages of vision is to suppress the attention-grabbing aspects of objects moving around what-

ever it is you want to pay attention to. See Y. Chen, S. Martinez-Conde, S. L. Macknik, Y. Bareshpolova, H. A. Swadlow, and J.-M. Alonso (2008), "Task difficulty modulates the activity of specific neuronal populations in primary visual cortex," *Nature Neuroscience* 11: 974–82.

69 Arturo de Ascanio: A. Ascanio, *The Magic of Ascanio*, vol. 1, trans. R. B. Etcheberry (self-published, 2007).

70 Nobel laureate Eric Kandel: E. Kandel, *In Search of Memory: The Emergence of a New Science of Mind* (New York: W.W. Norton, 2007).

74 Ethological studies: One evolutionary advantage of having a spotlight of attention dissociated from your center of gaze is that it enhances your ability to deceive others. Having a roving spotlight of attention that can point away from your direction of gaze allows you to hide what you are paying attention to (a potential food source, a desirable mate) from competitors. Marc Hauser at Harvard University has shown that monkeys will intentionally look away from hidden food sources in order to mislead other monkeys away from their stash. See M. D. Hauser (1992), "Costs of deception: Cheaters are punished in rhesus monkeys (Macaca mulatta)," *PNAS* 89(24): 12137–39. The cost of this system is that attending away from the fovea is, by definition, attending to low resolution information. Therefore, hiding your secret interests from those around you must convey an important adaptive edge.

74 In this sense, both macaques: Many other species use deception to maximize survival and reproductive success. Some birds will feign having a broken wing to lure a predator away from the nest: a form of misdirection. Such pretense of weakness is an old strategy in human warfare. Sun Tzu wrote in *The Art of War* more than two thousand years ago: "All warfare is based on deception. Hence, when able to attack, we must seem unable; when using our forces, we must seem inactive; when we are near, we must make the enemy believe we are far away; when far away, we must make him believe we are near. Hold out baits to entice the enemy. Feign disorder, and crush him." Other animals rely on camouflage and mimicry for deceptive purposes: some nonpoisonous butterflies evolved the same wing patterns as poisonous species, giving them the advantage of warning off predatory birds.

5. The Gorilla in Your Midst

77 To overcome adaptation: S. Martinez-Conde and S. L. Macknik (2007), "Windows on the mind," *Scientific American* 297: 56–63; S. Martinez-Conde, S. L. Macknik, X. G. Troncoso, and T. Dyar (2006), "Microsaccades counteract visual fading during fixation," *Neuron* 49: 297–305.

79 You cannot predict: For a more in-depth discussion of these ideas, see S. Martinez-Conde and S. L. Macknik (2008), "Magic and the brain," *Scientific American* 299: 72–79; S. L. Macknik, M. King, J. Randi, A. Robbins, Teller, J. Thompson, and S. Martinez-Conde (2008), "Attention and awareness in stage magic: Turning tricks into research," *Nature Reviews Neuroscience* 9: 871–79.

79 To describe these methods: Macknik et al., "Attention and awareness in stage magic."

80 Cognitive neuroscientists: A. Mack and I. Rock, *Inattentional Blindness* (Cambridge, Mass.: MIT Press, 1998).

80 Can You Keep Us From Reading Your Mind?: From Martinez-Conde and Macknik "Magic and the brain."

82 Tamariz uses inattentional blindness: Details can be found in his instructional masterpiece, *The Five Points of Magic.*

85 Our own research: Y. Chen, S. Martinez-Conde, S. L. Macknik, Y. Bereshpolova, H. A. Swadlow, and J. M. Alonso (2008), "Task difficulty modulates the activity of specific neuronal populations in primary visual cortex," *Nature Neuroscience* 11: 974–82.

85 The Gorilla in Our Midst experiment: For a wonderful and very entertaining in-depth look at this and related effects, see Chabris and Simons's new book, *The Invisible Gorilla* (New York: Crown Archetype, 2010).

85 In 2006, Daniel Memmert: D. Memmert (2006), "The effects of eye movements, age, and expertise on inattentional blindness," *Consciousness and Cognition* 15: 620–27.

86 Inattentional blindness: I. E. Hyman Jr., M. Boss, B. M. Wise, K. E. McKenzie, and J. M. Caggiano (2010), "Did you see the unicycling clown? Inattentional blindness while walking and talking on a cell phone," *Applied Cognitive Psychology* 24: 597–607.

87 Another of our colleagues: C. Rosen (2008), "The myth of multitasking," *New Atlantis: A Journal of Technology and Society* 20: 105–10.

91 In one version: D. J. Simons and D. T. Levin (1998), "Failure to detect changes to people during a real-world interaction," *Psychonomic Bulletin and Review* 5: 644–49. See also C. F. Chabris and D. J. Simons, *The Invisible Gorilla* (New York: Crown Archetype, 2010).

91 The experiment has been replicated many times: The British mentalist and magician Derren Brown loves change blindness and has made several video clips of the trick in London settings, based on the original Simons videos.

95 Slow or gradual changes: Chabris and Simons, *The Invisible Gorilla.*

6. The Ventriloquist's Secret

100 Senses not only interact: This research was carried out by Charles Spence, head of the Crossmodal Research Laboratory based at the Department of Experimental Psychology, Oxford University (http://www.psy.ox.ac.uk/xmodal/default.htm). He is interested in how people perceive the world around them—in particular, how our brains manage to process the information from each of our senses (smell, taste, sight, hearing, and touch) to form the extraordinarily rich multisensory experiences that fill our daily lives. He currently works on problems associated with the design of foods that maximally stimulate the senses, and with the effect of the indoor environment on mood, well-being, and performance.

101 The same goes for skin and sound: By mixing audio with the tactile sense of airflow, researchers at the University of British Columbia in Vancouver—linguistics professor Bryan Gick and his student Donald Derrick—found that perception of certain sounds relies, in part, on being able to feel these sounds. Their paper was published in *Nature*, November 26, 2009.

101 Your ears can also fool your eyes: L. Shams, Y. Kamitani, and S. Shimojo (2002), "Visual illusion induced by sound," *Cognitive Brain Research* 14: 147–52.

102 In the same vein: V. Jousmaki and R. Hari (1998), "Parchment-skin illusion: Sound-biased touch," *Current Biology* 8(6): R190.

102 How you feel the world can actually change how you see it: This research was carried out in the lab of Chris Moore at MIT and was published in the April 9, 2009,

online edition of *Current Biology*. Demos of the motion stimuli can be seen at http://web.mit.edu/~tkonkle/www/CrossmodalMAE.html.

102 And then there is the rubber hand illusion: M. Botvinick and J. Cohen (1998), "Rubber hands 'feel' touch that eyes see," *Nature* 391: 756.

102 The phenomenon is called *synesthesia*: R. E. Cytowic, *Synesthesia: A Union of the Senses*, 2nd ed. (Cambridge, Mass.: MIT Press, 2002); R. E. Cytowic, *The Man Who Tasted Shapes* (Cambridge, Mass.: MIT Press, 2003); R. E. Cytowic and D. M. Eagelman, *Wednesday Is Indigo Blue: Discovering the Brain of Synesthesia* (Cambridge, Mass.: MIT Press, 2009); J. E. Harrison, *Synaesthesia: Classic and Contemporary Readings* (Oxford, UK: Blackwell Publishing, 1996); A. N. Rich and J. B. Mattingley (2002), "Anomalous perception in synaesthesia: A cognitive neuroscience perspective," *Nature Reviews Neuroscience* 3(1): 43–52; E. M. Hubbard and V. S. Ramachandran (2005), "Neurocognitive mechanisms of synesthesia," *Neuron* 48(3): 509–20; J. Simner, C. Mulvenna, N Sagic, E. Tsakanikos, S. Witehrby, C. Fraser, K. Scott, and J. Ward (2006), "Synesthesia: The prevalence of atypical cross-modal experience," *Perception* 35: 1024–33.

102 Neuroscientists have identified at least fifty-four varieties of synesthesia: Caltech lecturer in computation and neural systems Melissa Saenz discovered this phenomenon quite by accident. She reported her findings, with neuroscientist Christof Koch, in the August 5, 2008, issue of *Current Biology*.

103 In mirror touch synesthesia: M. J. Banissy and J. Ward (2007), "Mirror-touch synesthesia is linked with empathy," *Nature Neuroscience* 10: 815–16.

103 As for the rest of us: The bouba kiki effect was first observed by the German-American psychologist Wolfgang Kohler. W. Kohler, *Gestalt Psychology* (New York: Liveright, 1929).

106 Have you ever driven a cat crazy: B. E. Stein, M. A. Meredith, W. S. Honeycutt, and L. McDade (1989), "Behavioral indices of multisensory integration: Orientation to visual cues is affected by auditory stimuli," *Journal of Cognitive Neuroscience* 1: 12–24.

106 *feature integration theory*: A. Treisman and G. Gelade (1980), "A feature-integration theory of attention," *Cognitive Psychology* 12(1): 97–136.

108 The concept was first: P. M. Roget (1825), "Explanation of an optical deception in the appearance of the spokes of a wheel seen through vertical apertures," *Philosophical Transactions of the Royal Society of London* 115: 131–40; S. L. Macknik (2006), "Flicker fusion," http://www.scholarpedia.org/article/Flicker_fusion.

109 Max Wertheimer . . . and Hugo Munsterberg: M. Wertheimer, *Drei Abhandlungen zur Gestalttheorie* (Erlangen, Germany: Philosophische Akademie, 1925); H. Munsterberg, *The Photoplay: A Psychological Study* (New York: D. Appelton and Co., 1916).

109 The dumbstruck editor sent: A. R. Luria and J. Bruner, *The Mind of a Mnemonist: A Little Book About a Vast Memory* (Cambridge, Mass.: Harvard University Press, 1987).

7. The Indian Rope Trick

111 According to Teller: Teller wrote his review in the Sunday *New York Times Book Review*, February 13, 2005. See http://www.nytimes.com/2005/02/13/books/review/13TELLERL.html.

117 False memories can be devastating: E. F. Loftus, *Eyewitness Testimony* (Cambridge, Mass.: Harvard University Press, 1996); E. F. Loftus and J. E. Pickrell (1995), "The formation of false memories," *Psychiatric Annals* 25(12): 720–25.

118 In one example: E. F. Loftus, "Made in Memory: Distortions in Memory after Misleading Communications," in G. Bower, ed., *The Psychology of Learning and Motivation,* vol. 30, *Advances in Research and Theory* (San Diego: Academic Press, 1993), 187–215.

118 In another classic experiment: E. F. Loftus and J. C. Palmer (1974), "Reconstruction of automobile destruction: An example of the interaction between language and memory," *Journal of Verbal Learning and Verbal Behavior* 13: 585–89.

120 Nader demonstrated that: O. Hardt and K. Nader (2009), "A single standard for memory: The case for reconsolidation," *Nature Reviews Neuroscience* 10(3): 224–34.

120 "Flashbulb memories": K. Nader (2003), "Memory traces unbound," *Trends in Neurosciences* 26(2): 65–72.

130 In an article for *Slate*: Joshua Foer, "Forget Me Not," Slate.com., March 16, 2005 (http://www.slate.com/id/2114925).

8. Expectation and Assumption

139 Theory of False Solutions: J. Tamariz, *The Magic Way* (Madrid, Spain: Frakson Books, 1988).

144 Eric Kandel: E. Kandel, *In Search of Memory: The Emergence of a New Science of Mind* (New York: W.W. Norton, 2007).

145 First, a visual region of your brain: R. A. Andersen and C. A. Buneo (2002), "Intentional maps in posterior parietal cortex," *Annual Review of Neuroscience* 25: 189–220.

146 Gustav Kuhn, a psychologist and magician: Gustav Kuhn and Micahel F. Land, "There's more to magic than meets the eye," *Current Biology* 16(22): 950–51.

147 If so, the neural correlate: J. A. Assad and J. H. Maunsell (1995), "Neuronal correlates of inferred motion in primate posterior parietal cortex," *Nature* 373: 518–21.

147 Subjects were asked to read a list of words: Study described in *Blink* by Malcolm Gladwell (Boston: Little, Brown, 2005).

148 Being reminded of their gender: Gladwell, *Blink.*

148 Half of the participants in another study: "Johan C. Karremans, Wolfgang Stroebe, and Jasper Claus, "Beyond Vicary's fantasties: The impact of subliminal priming and brand choice," *Journal of Experimental Social Psychology* 42(6): 792–98.

148 Advertisers use priming: J. L. Harris, J. A. Bargh, and K. D. Brownell (2009), "Priming effects of television food advertising on eating behavior," *Health Psychology* 28(4): 404–13.

149 There are certainly other contributors: Most magicians wouldn't perform this particular version of the trick onstage because it's not completely fail-safe. We include it here to illustrate priming in magic.

150 Signal detection theory: D. M. Green and J. A. Swets, *Signal Detection Theory and Psychophysics* (New York: Wiley, 1966).

151 Keith Payne: B. K. Payne (2001), "Prejudice and perception: The role of automatic and controlled processes in misperceiving a weapon," *Journal of Personality and Social Psychology* 81: 181–92.

153 Such questions raise a deeper quandary: J. Piaget, *The Origins of Intelligence in Children* (New York: International University Press, 1952); J. Piaget, *The Moral Judgment of the Child* (London: Kegan, Paul, Trench, Trubner and Co., 1932).

154 Elizabeth Spelke, a developmental psychologist: E. S. Spelke (1990), "Principles of object perception," *Cognitive Science* 14(1): 29–56.

154 Such research also shows that infants have: For a good review, see Laura Kotovsky and Renée Baillargeon, "The development of calibration-based reasoning about collision events in young infants," *Cognition* 67(3): 311–51.

155 He notes that infants: See www.cmu.edu/cmnews/030625/03625_cognition.html.

156 The famous Sally-Ann test: H. Wimmer and J. Perner (1983), "Beliefs about beliefs: Representation and constraining function of wrong beliefs in young children's understanding of deception," *Cognition* 13: 103–28.

157 "Adults can follow directions": J. Columbo, "Visual Attention in Infancy: Process and Product in Early Cognitive Development," in Alison Gopnik, *The Philosophical Baby* (New York: Farrar Straus and Giroux, 2009).

157 In an experiment by John Hagen: J. W. Hagen and G. H. Hale, "The Development of Attention in Children," in A. D. Pick, ed., *Minnesota Symposia on Child Psychology* (Minneapolis: University of Minnesota Press, 1973).

158 Silly Billy: D. Kaye, *Seriously Silly: How to Entertain Children with Magic and Comedy* (Washington, D.C.: Kaufman & Co., 2005).

9. May the Force Be with You

168 The effect is astounding: The mathematical explanation for this trick can be found at www.numericana.com.magic.htm.

172 Here is Dr. Anna Berti: A. Berti, G. Bottini, M. Gandola, L. Pia, N. Smania, A. Stracciari, I. Castiglioni, G. Vallar, and E. Paulesu (2005), "Shared cortical anatomy for motor awareness and motor control," *Science* 309: 488–91.

173 *choice blindness*: P. Johansson, L. Hall, S. Sikstrom, and A. Olsson (2005), "Failure to detect mismatches between intention and outcome in a simple decision task," *Science* 310: 116–19.

173 Johansson explains that their experiments were inspired: See Richard E. Nisbett and Timothy D. Wilson (1977), "Telling more than we can know: Verbal reports on mental processes," *Psychological Review* 8: 231–59, and http://en.wikipedia.org/wiki/introspection_illusion.

175 In a follow-up experiment: The studies about preferences for jam and tea and the magic questionnaire have been submitted for publication. For the latest updates, see Petter Johansson's Web site, http://www.lucs.lu.se/petter.johansson/.

176 Again, the results show that a majority of the participants are blind: See Johansson and Hall's Web site (http://www.lucs.lu.se/projects/choiceblindness) and a YouTube video (http://www.youtube.com/watch?v=WBO03PngZPU).

178 Our colleagues Apollo Robbins . . . and . . . Eric Mead: See video at http://www.sfn.org/index.aspx?pagename=am2009_highlights.

180 In the 1970s: The original paper is B. Libet, C. A. Gleason, E. W. Wright, and D. K. Pearl (1983), "Time of conscious intention to act in relation to onset of cerebreal activity (readiness-potential)," *Brain* 106: 623–42.

181 A politician in the mayor's office: See http://www.youtube.com/watch?v=-o_pYTOodu4.

183 John-Dylan Haynes: C. S. Soon, M. Brass, H. J. Heinze, and J. D. Haynes (2008), "Unconscious determinants of free decisions in the human brain," *Nature Neuroscience* 11(5): 543–45.

186 We may believe that they are connected to free will: See the work of John Bargh at Yale, http://www.yale.edu/psychology/FacInfo/Bargh.html.

186 The illusion of the magic self cannot be easily suppressed: One of the more interesting findings in the free will literature is that when people believe, or are led to believe, that free will is an illusion, they may become more antisocial. Kathleen Vohs from the University of Minnesota and Jonathan Schooler from the University of British Columbia brought thirty students into their lab for a study that was supposedly about mental arithmetic. The students were asked to calculate answers to twenty simple math problems in their heads. Before taking the test, however, half read this passage from Francis Crick's book *The Astonishing Hypothesis*: "'You,' your joys and your sorrows, your memories and your ambitions, your sense of personal identity and free will, are in fact no more than the behavior of a vast assembly of nerve cells and their associated molecules. Who you are is nothing but a pack of neurons . . . although we appear to have free will, in fact, our choices have already been predetermined for us and we cannot change that." The other fifteen students read a different passage that did not mention free will. Later, given the chance, the students who read the more neutral passage cheated less than the group who had read that free will is an illusion. For a discussion of this experiment, see http://www.scientificamerican.com/article.cfm?id=scientists-say-free-will-probably-d-2010-04-06. See also D. M. Wegner, *The Illusion of Conscious Will* (Cambridge, Mass.: MIT Press, 2002).

186 Moreover, many philosophers and scientists argue: See Daniel C. Dennett, *Freedom Evolves* (New York: Viking Penguin, 2003).

10. Why Magic Wands Work

189 the *ideomotor effect*: *Dowsing* is a type of divination used in attempts to locate groundwater, buried metals or ores, gems, oil, graves, and other objects beneath the surface of the earth. The dowser holds a Y-shaped rod that magically "bends" when the dowser is standing over the sought target. *Automatic writing* is the process of writing that does not stem from conscious thought; it is done by people in a trance state. *Facilitated communication* is a process by which a facilitator supports the hand or arm of an impaired person—often someone with autism—to help them write and communicate. All three practices are examples of the ideomotor effect.

194 But nobody ever fails: The season two finale of the TV show *Lost* revealed that pushing the button indeed discharged an electromagnetic field that would otherwise continue to grow until ultimately causing the end of the world. Thus pushing the button to avert world-scale destruction turned out to be a real cause-effect relationship rather than an illusory correlation. But at the beginning of the season, when the characters resign themselves to push the apparently ineffectual button every 108 minutes, they have no factual data that the correlation is real.

194 a team of cognitive neuroscientists: B. A. Parris, G. Kuhn, G. A. Mizon, A. Benattayallah, and T. L. Hodgson (2009), "Imaging the impossible: An fMRI study of impossible causal relationships in magic tricks," *Neuroimage* 45(3): 1033–39.

194 Your implicit system of knowledge of cause and effect: See *Neuroimage* 45(3): 1033–39.

195 ACC, detects conflict: M. M. Botvinick, T. S. Braver, D. M. Barch, C. S. Carter, and J. D. Cohen (2001), "Conflict monitoring and cognitive control," *Psychological Review* 108: 624–52.

199 If the mentalist never misses: Credit for this observation goes to Magic Tony.

207 Some people looked: A. Raz, T. Shapiro, J. Fan, and M. I. Posner (2002), "Hypnotic suggestion and the modulation of Stroop interference," *Archives of General Psychiatry* 59: 1155–161.

208 Sixteen people . . . came into Raz's lab: *Cortex* 44(10): 1336–41.

209 COMT may confer susceptibility: P. Lichtenberg, R. Bachner-Melman, I. Gritsenko, and R. P. Ebstein (2000), "Exploratory association study between catechol-O-methyltransferase (COMT) high/low enzyme activity polymorphism and hypnotizability," *American Journal of Medical Genetics* 96(6): 771–74.

210 "In fact, I was the victim": See Zak's blog: http://www.psychologytoday.com/blog/the-moral-molecule.

214 Oxytocin causes us to empathize with others: A recent study indicates that oxytocin is not all touchy-feely. Experimental subjects who inhaled oxytocin while playing a competitive game in the laboratory experienced stronger feelings of envy and gloating than subjects exposed to a placebo. The researchers speculated that oxytocin might intensify social emotions in general, leading to generosity and trust in positive situations and to envy and gloating in competitive scenarios. See http://www.scientificamerican.com/article.cfm?id=oxytocin-hormone.

214 With a magician, you know you're being scammed: See Paul Zak's blog in *Psychology Today*, November 13, 2008. His book *The Moral Molecule* will be published in 2012 by Dutton.

11. The Magic Castle

222 Professional pianists (and magicians!): A. Pascual-Leone, D. Nguyet, L. G. Cohen et al. (1995), "Modulation of muscle responses evoked by transcranial magnetic stimulation during the acquisition of new fine motor skills," *Journal of Neurophysiology* 74: 1037–45; A. Pascual-Leone (2006), "The brain that plays music and is changed by it," *Annals of the New York Academy of Sciences* 930: 315–29.

222 Here is one more: S. Blakeslee and M. Blakeslee, *The Body Has a Mind of Its Own* (New York: Random House, 2007).

222 The dance has become part of his being: See Blakeslee and Blakeslee, *The Body Has a Mind of Its Own*. See http://www.youtube.com/watch?v=-X0AamE1Bxs for an idea of what happens if you learn the samba as a baby.

225 They are so good: M. Natter and F. Phillips (2008), "Deceptive biological motion: Understanding illusionary movements," *Journal of Vision* 8(6): 1052.

232 The sheikh nearly fainted: In 1856, Louis-Napoleon asked Robert-Houdin to convince certain Arab chieftains that the French war machine had magical powers. Religious tribal leaders called marabouts, who used magic to control their followers, had advised their chieftains to break with the French. Napoleon wanted Robert-Houdin

to convince the Arabs that French magic was stronger than Arab magic—thus avoiding a war in Algeria. One evening in a stifling hot theater in Algiers, Robert-Houdin demonstrated his powers to the assembled chieftains. He produced a cannonball from a hat. He passed around an inexhaustible bottle that dispensed hot coffee. But his pièce de résistance was issued as a challenge: "I can deprive the most powerful man of his strength and restore it at my will," said the French magician. "Anyone who thinks himself strong enough to try to experiment may draw near me." A muscular man approached. "Are you very strong?" "Oh, yes." "Are you sure you will always remain so?" "Quite sure," the man replied.

"You are mistaken," said Robert-Houdin, "for in an instant I will rob you of your strength and you shall become as a little child." Pointing to a small wooden box, he said, "Lift up this box." The man lifted the box and laughed. "Is that all?"

Robert-Houdin said "Wait!" and then, making an imposing gesture, "Behold." He waved his magic wand. "Now you are weaker than a woman. Try to lift the box." The man tried. He pulled with all his might. Sweat poured down his face. He tried to rip the box apart, to no avail. You see, the box contained a powerful electromagnet, which exerted a force unknown to the marabouts. Robert-Houdin then delivered an electric shock to the man, who ran screaming off the stage.

With this display of French supernatural power, the rebellion was put down.

235 In 2007, a retired CIA officer: H. Keith Melton and Robert Wallace, *The Official CIA Manual of Trickery and Deception* (New York: William Morrow, 2009).

12. Will the Magic Go Away?

250 Your mirror neuron system: S. Blakeslee and M. Blakeslee, *The Body Has a Mind of Its Own* (New York: Random House, 2007).

250 The same goes for athletics: B. Calvo-Merino, D. E. Glaser, J. Grezes, R. E. Passingham, and P. Haggard (2005), "Action observation and acquired motor skills: An fMRI study with expert dancers," *Cerebral Cortex* 15(8): 1243–49.

252 And it could work the other way around: Magicians are beginning to use in their stage acts perceptual effects originally designed for scientific experiments. Derren Brown and Penn & Teller execute change blindness routines that are firmly rooted in the cognitive sciences. Teller says of the change blindness routine in the Penn & Teller act, "The idea came straight from science. We thought it would be fun to show how bad they are at noticing stuff" (J. Lehrer, "Magic and the brain: Teller reveals the neuroscience of illusion," Wired.com, April 20, 2009).

255 You might wonder: In fact, a few of our colleagues, led by Peter Lamont, a lecturer in psychology at the University of Edinburgh, have made this suggestion.

257 We believe it also determines: Y. Chen, S. Martinez-Conde et al. (2008), "Task difficulty modulates the activity of specific neuronal populations in primary visual cortex," *Nature Neuroscience* 11(8): 974–82.

ACKNOWLEDGMENTS

This book is the product of a yearlong collaboration between the two of us and Sandy Blakeslee, who is in our opinion the most gifted and superlative neuroscience writer in the world. Sandy not only took our content and made it beautiful, she also brought an enormous amount of neuroscientific knowledge to the table. Backed by her most recent book, *The Body Has a Mind of Its Own*, Sandy could easily teach courses on the neuroscience of motor control in the brain, at the graduate level. She certainly taught us about how to think about many of the magical effects we've been studying from the motor-control perspective. This book is so much stronger for her participation that we simply cannot thank her enough for her insights, her beautiful writing, and her patience with both our own dreary offerings and our crazy work schedule.

Matt Blakeslee, Sandy's son and cowriter on *The Body Has a Mind of Its Own*, is also a brilliant writer with a vast knowledge of neuroscience. He helped us tremendously on creating the proposal and framing the story of the book.

Dozens of selfless magicians and scientific colleagues also offered their time and expertise freely and, often, with impossibly short deadlines. A year is a very short cycle to produce a book. In the lingo of publishers, they say that our book has been crashed. As such, every deadline was tight, every request for information or fact checking was rushed, and we are very grateful to everybody for pulling together to help us on such short notice.

Our thanks go out especially to the original group of magicians who taught us the importance of magic as a neuroscientific venture (in alphabetical order): Mac King, James Randi (The Amaz!ng Randi), Apollo Robbins, Teller, and Johnny Thompson (The Great Tomsoni). They've guided us through this project throughout, including proofreading the magical parts of the book for accuracy. Any errors are ours, certainly, and we must admit that any brilliant insights are probably theirs.

Anthony Barnhart (Magic Tony), the outstanding young magician here in Phoenix who had the onerous task of actually teaching us to do magic with our hands, is a saint. Reading books like this one can help you understand why magic is cool intellectually. But learning to do tricks well enough to actually fool someone will take you to another level altogether. Tony also contributed many important scientific insights. So he gets our deepest thanks as well.

Dozens of other magicians have also contributed their time and insights. They include (in alphabetical order): Francesc Amílcar, Jerry Andrus, Mago Antón, Luis Boyano, Eugene Burger, Jack Devlin, Ava Do, Paul Draper, Jesús Etcheverry, Miguel Ángel Gea, Roberto Giobbi, Larry Hass, Danny Hillis, Joshua Jay, Penn Jillette, Isaac Jurado (Mago Isaac), Bill Kalush, David Kaye (Silly Billy), Jason Latimer, Patrick Martin, Max Maven, Jeff McBride, Eric Mead, Tom Meseroll (Magus, the Martial Magician), Harry Monti, Luis Piedrahita, Shoot Ogawa, Gabi Pareras, Kiko Pastur, Adam Rubin, Jay Sankey, Victoria Skye, Scotto Smith, Jamy Ian Swiss, Juan Tamariz, Ángel Vicente, Timothy Vient, Allen Waters, Michael Weber, and Richard Wiseman.

Many of our academic colleagues advised us, with our thanks: José-Manuel Alonso, Dan Ariely, Mahzarin Banaji, Dan Dennett,

David Eagleman, Lars Hall, Petter Johansson, Christof Koch, Gustav Kuhn, Joseph Ledoux, Luis Martínez, Michael Natter, Flip Phillips, Dan Simons, Benjamin Tatler, Paul Zak, and Stuart Zola.

We are very grateful to the various funding organizations that have supported our research or our other scientific endeavors that eventually led to this book: the Barrow Neurological Foundation, the Mind Science Foundation, the National Science Foundation, the National Institutes of Health, Science Foundation Arizona, and the Arizona Biomedical Research Commission. Special thanks go to Joseph Dial of the Mind Science Foundation for spurring us along for so many years.

Phil Pomeroy and Lynne Reaves at the Barrow Neurological Institute have been wonderful to us and the strongest possible supporters and promoters. Thank you.

We are indebted to María Dolores Ruiz and Guillermo Santamaría of Fundació "la Caixa," and also Marcos Pérez of Casa de las Ciencias, for bringing us together with so many colleagues in Spain.

Tom Carew, the 2009 president of the Society for Neuroscience, has been a wonderful friend. He made neuromagic a central topic for the annual conference of the SfN, the largest academic conference in the world. More than seven thousand neuroscientists attended the session.

Some of the illustrations were made at our request, and the terrific artists who made them at the Barrow Neurological Institute include: Mark Schornak, Marie Clarkson, Mike Hickman, and Jorge Otero-Millan.

The marine and navy officers who helped us understand the issues surrounding situational awareness in military aviation were very generous with their time: Ellis Gayles, Michael Prevost, Paul Gosden, and Vincent Bertucci.

This book would have never happened if it wasn't for Mariette DiChristina, the editor in chief of *Scientific American*, who kindly sought dozens of our contributions for the Scientific American family of publications. Without her very kind encouragement, friendship, and thoughtful editing, we wouldn't have dared to attempt to write for the public. We're also grateful to Peter Brown for doing such a great job in editing our "Magic in the Brain" *Scientific American* cover story in December 2008. David Dobbs, author of *Reef Madness: Charles Darwin, Alexander Agassiz, and the Meaning of Coral*; Jonah Lehrer, author of *Proust Was a Neuroscientist* and *How We Decide*; and Pulitzer Prize winner Gareth Cook have all served as editors of our column "The Neuroscience of Illusion" in the Mind Matters section on ScientificAmerican.com, and we're grateful to all of them for helping us so much to learn to write for the public.

We also thank George Johnson and Ben Carey, of the *New York Times*, who covered our magic symposium and brought it to many people's attention in 2007 and 2008. They have also been terrific and insightful advisers since.

The book has been guided expertly through the entire publication labyrinth by Gillian Blake, executive editor at Henry Holt and Company, and her assistant Allison McElgunn. Their editing, in addition to that of Emily DeHuff, served to make the final product of this book exactly what we had hoped for and more.

Orchestrating the entire enterprise on the publishing side has been Jim Levine, our literary agent at Levine|Greenberg in New York City. He and his team including Kerry Sparks and Elizabeth Fisher have made the business end of this project seem easy. Pure magic on their part, we're sure.

Finally we want to thank our families. Susana's sister, Carolina Martínez-Conde García, helped us understand the nature of the gambler's fallacy from the point of view of a croupier and she also digitized all of their grandfather Enrique's diaries so that Susana could use them in chapter 10. Susana's godmother, María de la Cruz García Blanco, helped us understand the story of her father, Enrique, and his role in the Spanish Civil War. Our two little boys, Iago and Brais, were both born after we started working with magicians and they've been immersed for their entire lives. The project has taken us all over the world, to most of the continents, and to more than a dozen countries. We dragged them along with us the whole way. Our mothers, Sarah Macknik and Laura Cruz García Blanco, were wonderful and often traveled with us to help take care of our children, but it's really the boys who playfully suffered the brunt of the inconvenience to their young lives. Iago, at three years of age, has his own gold member frequent flier card with US Airways. Playing in a different hotel every week, seeking out children to play with in anonymous restaurants—they've been real troupers and we're so grateful to them for the magic they bring to our lives.

INDEX

Page numbers in *italics* refer to boxes and illustrations.